"It is well recognised that a collaborative relatior
arm's length adversarial one. Unfortunately, there
failed collaborative relationships that failed to del
authors of this book quite rightly identify that col ... organisational partners takes more than good intentions. It takes a well-planned and communicated approach that is ideally jointly developed between the parties. In this book, you will find a step-by-step guide with worksheets and tools that you and your intended partner can use to develop a joint plan, an execution strategy, and an ongoing relationship management process to keep you on track. It also recognizes that the best time to work on your exit plan from the relationship is at its beginning when the relationship is at its strongest. I would recommend this book to anyone who is either engaged in developing or already involved in collaborative relationships as a tool to deliver on the promise that your relationship can deliver."

– Andrew Downard, Ph.D., CEO, Ci-Advisory, Australia

"This book is an invaluable resource for anybody interested in managing business relationships. Humphries and McComie present an authoritative account of how collaborative relationships work and why they don't work. The practical tools, such as worksheets and checklists, provide a hands-on structure to ensure successful relationship management."

– Dr. Carlos Mena, Nike Professor of Supply Chain Management, Portland State University, USA

"The concept that collaboration delivers better results is not new, but focus on short-term wins frequently stands in the way. The global pandemic has provided a wake-up call, an appreciation that 'we are in this together.' With more than 60% of commercial executives now pushing for collaborative relationships, this practical 'how-to' guide could not be more timely."

– Tim Cummins, President, World Commerce and Contracting; Professor, International Commercial and Contract Management, University of Leeds School of Law

"The successive shocks of the pandemic, microchip supply and now emerging inflationary pressures have challenged supply chains more than at any time in the last 20 years. Resilience has now become a key distinguishing factor between organizations, and those that have built the strongest collaborative relationships and networks are best placed to maximize profitability in 2022 and beyond."

– Mike Reilly MBE, CEO – ETher NDE Ltd., Baugh & Weedon Ltd., Alban NDE LLC

Implementing and Managing Collaborative Relationships

Implementing and Managing Collaborative Relationships

A Practical Guide for Managers

Andrew Humphries and Linda McComie

Routledge
Taylor & Francis Group
A PRODUCTIVITY PRESS BOOK

First published 2022
by Routledge
605 Third Avenue, New York, NY 10158

and by Routledge
4 Park Square, Milton Park, Abingdon, Oxon, OX14 4RN

Routledge is an imprint of the Taylor & Francis Group, an informa business

ISBN: 9781032117508 (hbk)
ISBN: 9781032117386 (pbk)
ISBN: 9781003221340 (ebk)

DOI: 10.4324/9781003221340

Typeset in Garamond
by Deanta Global Publishing Services, Chennai, India

Contents

About the Authors

Andrew Humphries. Andrew is CEO of SCCI Ltd, a company that specializes in performance improvement within complex, commercial, and government relationships. Over the last 20 years, he has worked with a wide variety of organizations in Europe, Asia, and Australia using PartnerLink, a scientific process that clearly identifies partnering performance drivers and opportunities for improvement. As a result, he has enabled many to achieve significantly enhanced collaborative effectiveness and bottom-line benefits. He retired from the UK RAF in 2004 as Head of Policy for Defense Aviation Logistics. He has over 35 years of experience as a practical military logistics manager and director. Andrew gained his PhD from Cranfield School of Management, UK, in 2003 and has published widely. His books *Strategic Alliances and Marketing Partnerships* (Kogan Page 2009), *Collaborative Change* (2010), and *Enterprise Relationship Management* (Gower 2015) demonstrate the importance of applying a management discipline to ensure the success of collaborative relationships, alliances, and partnerships.

Linda McComie. Linda is the Director and co-founder with Andrew Humphries of SCCI Ltd. Over the last 20 years, she has piloted and co-developed the SCCI toolset with Andrew. She has been actively promoting the use and continuous development of collaborative working with both managers and business schools. Her work with SMEs has been particularly successful, where client's organizations grew by 33% year on year over a three-year monitored period. She is a former Senior Manager in the UK Finance, Manufacturing and Employment sectors and has worked extensively with blue chip companies to improve the performance of their alliances and collaborative partnerships. Linda gained her MBA from the Open University Business School in 1994. She is the co-author of the SCCI Simple Guide to Collaboration series of books (Taylor & Francis).

SCCI Ltd is a small, innovative, knowledge-based enterprise. It supports companies around the globe to transform their business relationships and alliances into more effective and efficient revenue-generating operations. SCCI is greatly concerned that traditional customer/supplier management approaches deny companies the benefits of leveraging the wider opportunities of alliances. Historically, companies that systematically manage their relationships are able to generate up to four times the market wealth. SCCI provides a key enabler to realizing these benefits: high-quality objective performance metrics and diagnostics for both governance and management. SCCI's clients come from a range of sectors including defense, rail, construction, electronics, finance, manufacturing, automotive, IT, electronics, oil and gas, agriculture, food, outsourcing, facilities management, and retail fashion.

Foreword

Increasingly it is being recognized that companies can no longer sustain competitive advantage without effective collaboration with multiple stakeholders. These stakeholders may include suppliers, customers, government, NGOs (non-governmental organizations), and joint venture partners, just to name a few. Each can have a significant impact on the success or otherwise of a business. The concept of Enterprise Relationship Management (ERM) defined as "the management process for coordinating business activities to the success of a joint or multi-party endeavor" is essential for the success of increasing numbers of businesses today.

The recognition that 'Competition is no longer between individual organizations but the supply chains they are part of' and 'Supply chain management is the management of up-stream and down-stream relationships with all stakeholders including customers and suppliers in order to create enhanced value in the marketplace at less cost to the supply chain as a whole' emphasizes the need for effective relationship management with supply chain partners to remain competitive.

'Value', those things customers and stakeholders perceive as important and beneficial, is increasingly created through relationships, both internal to the organization and external to it. The increased awareness of the 'triple bottom line' which focuses an organization on economic value (Profits), societal value (People), and environmental value (Planet) has driven organizations to build and maintain relationships in new ways. Traditional procurement philosophy of identifying suppliers that can be 'leveraged' by driving costs and margins down when viewed through the lens of the triple bottom line is seen as an increasingly archaic strategy that in many instances destroys rather than creates value. The shift in recent years toward procurement for resilience rather than cost and the recognition that innovation is often created by the relationships you create with suppliers requires proactive relationship management rather than re-active relationship management to succeed.

It is, however, worrying that for many businesses, relationships are managed and developed in an unplanned, often 'accidental', manner with the individuals responsible for supporting business-critical relationships receiving little guidance and support. Our education system often emphasizes the 'Technical Skills', the IQ, of operating within a business with little time given to developing the 'Relational Skills', the EQ (Emotional Intelligence). This, it can be argued, increases the risk profile of the organization and its supply chain with poor relationship management both internally and externally, even resulting in a business failing. Many businesses have no formal process for developing, managing, and exiting relationships. They allocate little resource including people, information systems, and even infrastructure to relationship management. Few organizations have an enterprise relationship management strategy; few have documented processes and can verbalize how relationships are created and maintained. Even fewer can describe how relationships are exited in a way that all parties remain whole through 'conscious uncoupling' to allow relationships to reform again when new value opportunities are identified. It is not uncommon that when managers are asked what documentation they have which describes the approach to the relationship, they produce an inflexible legal contract which outlines the consequences of failure!

In order to 'manage' relationships, goals need to be set and resources need to be mobilized effectively to achieve the goal. Measurement and monitoring becomes critical, and over the past 20 years, techniques have been developed to enable the measurement and monitoring of relationships from both soft and hard perspectives. Sadly few companies set 'relationship goals' and even fewer 'measure' and 'monitor' the rich resource of relationships they have.

In order to address the above issues and formalize an approach to developing and managing business relationships, both academics and practitioners have recognized the need for a more methodical approach to creating and managing collaborative relationships. It is recognized that all relationships are different, but there are common building blocks and approaches to their implementation and management. This is one of the very few books that provides a pragmatic holistic methodology that can be readily adapted for all business situations. It is not just a 'nice idea'. Its principles and techniques can be applied at all levels of relationships, from a micro (one to one) level, to macro (team to team) level, to the global level (enterprise to enterprise). Its implementation results in increased value for all stakeholders, and in my experience cost reduction can also occur simultaneously.

This book provides an excellent overview of the tools techniques and philosophy behind enterprise relationship management. It complements much of our work at Cranfield School of Management, UK, where taking academic knowledge and creating action in business is foundational. Particularly useful are the Worksheets at the end of each chapter that enable the reader to transfer the guidance into their own organization's relationships. Thus, it will provide an essential resource to enable a more formalized approach to business relationship management.

> During the COVID-19 pandemic, we have seen how those supply chain managers with the strongest network of relationships have coped better. Businesses tend to focus effectively on the basic management of inventories, KPIs, information systems and people – but appear to think relationships happen by accident. Collaboration and proactively managed relationships are the key to future resilience.

Richard Wilding OBE
Professor and Chair of Supply Chain Strategy,
Cranfield School of Management,
Cranfield, Bedford, MK43 0AL, UK
http://www.som.cranfield.ac.uk/lscm/

Chapter 1

Implementing and Managing Collaborative Relationships – An Introduction

The Purpose of This Guide and How to Use It

Our purpose is to show you how to put together a simple system of collaborative relationship management that you can tailor to the needs of your organization – whether you are an SME or a multi-national company, selecting a new partner or managing existing relationships. It will allow you to get the maximum value from your collaborative relationships, such as improved revenues, reduced costs, greater market share, specialized know-how, and better resilience in difficult conditions. For instance, uncertainty challenges the partners to adapt and capitalize on problems raised by unforeseeable events such as the recent global pandemic, ocean supply chain disruptions in the Suez Canal, local wars, and terrorist activities. Adaptability requires a high degree of coordination, cooperation, and collaboration by relationship partners. However, these require significant management effort and are critical to relationship survival.

> Whether you have a Supplier in Tianjin or 15 miles down the road, whether you have a Joint-Venture Partner in Madrid or New York, whether you have an OEM Customer in Loughborough or Sao Paulo, an unrelenting focus on developing and maintaining mutually successful medium and long-term business partnerships

DOI: 10.4324/9781003221340-1

through meaningful collaboration is critical to building a company in the 21st century. After all, with the complexities and competitiveness of today's business environment, none of us 'can do it all on our own', can we?

Mike Reilly, CEO, Ether NDE

Most companies will have well-established management systems; however, these are not usually optimized to manage beyond the organization's boundaries. A joint enterprise, working with partners, is completely different and therefore will need a new type of management system. There are significant differences in attitude between working on your own and working with others; it's no longer 'me', but it's 'us'. In a joint enterprise, the partners must learn how to earn each other's trust. Our purpose is to describe such a system and show you how to implement and successfully manage in a collaborative environment. As you get better at managing your relationships, you will find that inevitably your financial and other business returns will continue to improve as will your reputation as a company that others want to do business with.

Transactions or Collaboration?

There are two main types of relationships – those tending toward transactional and those tending toward collaborative. Transactional relationships are those based on the simple purchase or supply of a widely available product or service which is driven by price, quality, and time. Often transactional relationships become collaborative as trust develops over time. Conversely, collaborative relationships are those where the partners work together using their specialized resources innovatively to achieve aims and objectives that could not be realized on their own. Organizations must continually identify whether dealings with both prospective and existing suppliers are transactional or collaborative. In the latter situation, the managers must decide whether there is a compelling benefit to a closer relationship. Thus more often than not collaboration emerges rather than being created from scratch.

Collaborative Relationships

The most valuable of all business relationships are likely to be those that are collaborative, regardless of where they fit within the spectrum of relationship types. They could be buyer–seller relationships, supply chain partnerships, services marketing arrangements, facilities management, R&D, or any other relationship combination. They are likely to be strategically important and of long duration and, therefore, collaborative relationship management principles will apply.

However, managing them is often seen as a 'black art' where we don't know what to do, we don't know that we don't know what to do, it's somebody else's job, or we don't have the time or money to do it. These attitudes are prevalent, despite the high value (both bottom line and strategic) that is usually tied-up in the often long-standing formal and informal agreements between collaborating organizations of any size.

Achieving Collaborative Value

Collaborating organizations will bid for contracts that were previously out of reach, put products on the shelves at a fraction of the cost of their competitors', dream up a new product which will corner the market, and provide a super service package well beyond anything they have ever done before. Two years later when none of these objectives have been achieved, they have no idea why things have gone wrong. There is thus a gulf between these grandiose views of collaborative value and the ability of managers to translate them into achievable goals. Contrary to some traditional views that collaborative relationships can be established immediately simply by having the right partner selection process and signing the contract, it is clear that collaborative working, value creation, and capture develop in an increasingly productive cycle.

Traditionally, value (benefits) is seen as the achievement of increased profits, greater market share, and the fulfillment of project objectives. These benefits will be underpinned by operational imperatives of time, cost, quality, and delivery and are easy to quantify and track. However, there is more to value than this. A number of intangible factors will also come into play such as innovation, knowledge and skills capture, enhanced reputation and brand value, and satisfied stakeholders. More difficult to quantify, but equally important as an enabler, is relationship quality. This is characterized by harmony, trust, commitment, open communications, and a willingness to work together for the long term.

> The relationship has been absolutely outstanding. The trust established makes us want to do anything we can for the customer.
>
> **Project Director, Defense Supplier**

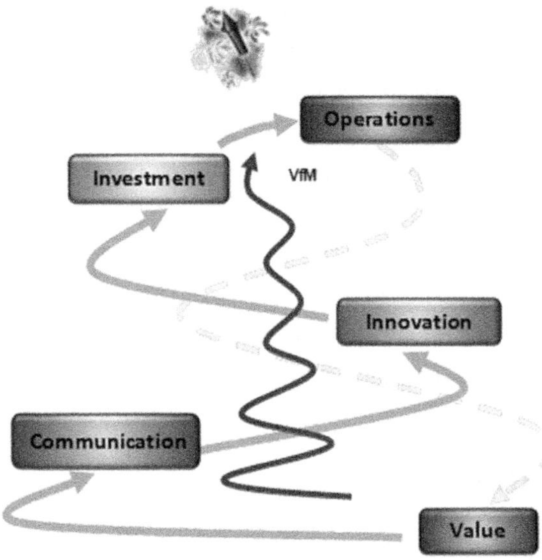

Figure 1.1 The virtuous relationship value cycle.

When companies start working collaboratively, they often begin with specific aims. These will take into account the known capabilities on both sides. As they move forward, further abilities will become evident. More open and in-depth communication will occur, creativity and innovation will be stimulated, and the partners will broaden their focus to include these and 'up their game'. Suddenly ideas that were vague dreams become achievable. These provide impetus for joint investment that enables the partners to seize unforeseen opportunities as they arise. It is this enhanced capability of the partnership that enables them to recognize and take advantage of potential collaborative value. It becomes a method of working that is capable of continuous value creation as the quality of the relationship increases. Therefore, the partners must expect that new value opportunities will emerge and be prepared to manage their creation and exploitation. Overall, it is important that benefit realization is tracked so that it sustains the impetus of the collaborative enterprise. Benefits encourage innovation which generates more value which can be reinvested in a virtuous cycle (Figure 1.1).

Relationship Management in Principle

Relationship management is one of the most important management systems that you can have, and as such, it should be clearly documented both in policy and in practice. Good relationship management has the capability to give you startling results, but, without it, full relationship value is unlikely to be achieved. Business tends to follow a standard 'Plan, Do, Check, Act' action cycle. Relationship management principles use a slightly different virtuous circle that acknowledges the joint nature of the endeavor (Figure 1.2).

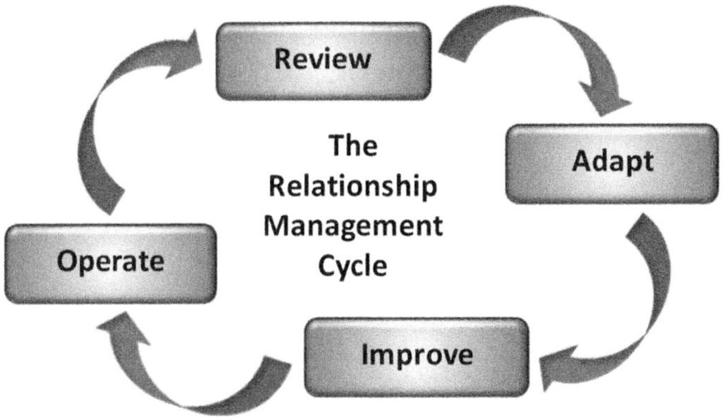

Figure 1.2 The relationship management cycle.

The partners first 'Review' the quality of their collaborative relationship and its achievements. They then 'Adapt' their way of working to keep their relationship in-step. Next, they agree 'Improvements' to their processes and behaviors. Finally, they 'Operate' together. In this way, continuous improvement is achieved and performance increases become inevitable.

Relationship Management in Practice

However, most relationships will miss opportunities to capitalize on their strengths because no one has an overall view of the joint enterprise. More often than not, the relationship management that exists is scattered among a number of individuals: the bosses at their annual game of golf, the salesman who is keen to keep the customer happy so he can land the follow-on deal, the commercial officer who manages the contract and sorts out the problems left by the salesman, and, finally, the administration, logistics, and operations people who have the job of making things happen. In the case of the micro business owner, it's tucked away in the back of his head while trying to juggle 20 things at once.

Many organizations do recognize the need to work more closely with partners, but their efforts miss the target. For example, they will take a 'gung-ho' attitude and remove governance arrangements which usually result in cost or time repercussions. They may re-badge their supplier management team as 'Vendor Managers' team which just perpetuates the old 'them and us' way of doing things. Some will see the adoption of standards as the way forward and fall into the trap of managing the standard rather than the relationship. They then wonder why this investment is not paying off.

> There is suspicion on all sides regarding the relationship. This is fueled by the customers' approach to management (micro-management) and the supplier's unwillingness to be open in discussion. Both parties are contributing to the poor performance of the contract.
>
> **Regional Director, IBM**

In the case of a supply chain, the end-to-end perspective is lost among the people who manage the individual links. This is particularly evident in the just-in-time situation where the links between the organizations have become 'mechanized'. In good times, these are very efficient; however, they are highly vulnerable to disruption because there is no relationship management in place to effectively overcome the risks.

The prospect of doing business across organizational boundaries within collaborative relationships is a daunting one. Fulfilling strategic needs through collaboration by selecting the right partner(s) and improving the performance of existing relationships will certainly only be fully successful if relationship management is effective. Continuous improvement will include diverse activities such as developing joint and outsourced processes (do you do your own accounting and website design?), rescuing failing partnerships, increasing the depth of collaboration with an existing partner to undertake a more challenging assignment, or improving the chances of partnering success at the 'pre-nuptial' stage.

> The benefit of our partnering arrangement is we are 'future proofed', uncertainty is removed, we can plan, and we can really focus on the customer.
>
> **CEO, Specialist Food Manufacturer**

A Target Operating Model for Managing Collaborative Enterprises

Much has been written on the subject of collaboration between organizations, the principles of trust, maturity matrices and where you are positioned within them, good practice tips, and secrets of success. But, none tell you how to do it in a simple and effective way. This book is all about a relationship management target operating model based on collaborative working principles which will enable you to achieve collaborative value and more specifically to:

- Critically scrutinize your existing partnerships and where needed, sharpen them up
- Efficiently set up a new relationship
- Run and continually improve joint operations
- Build management capability, recognize, and seize new opportunities
- Exit cleanly and learn positive lessons

Do as little or as much as you need to do, and the increase in business performance will repay your investment many times over.

The Implementation Journey

This book will cover the below topics in the following chapters:

Chapter 2 Enterprise Relationship Management (ERM) – A Target Operating Model for Collaboration: This describes the management process for coordinating all the business activities that are essential to the success of a joint/multi-party endeavor.

Chapter 3 The Relationship Manager – Clears Tall Buildings in One Bound: It is essential that each party has a person responsible for managing their involvement in the relationship. This chapter describes the role in detail and provides a specimen job specification.

Chapter 4 The Enterprise Relationship Management Plan (ERMP) – The 'Dashboard': This is the 'dashboard' that provides the 'instruments' and 'controls' to manage the relationship. It brings together the essential information necessary to set up the partnership, manage it, and coordinate all the key activities. It is the collaborative target operating model in use. We provide a detailed framework containing instructions for use.

Chapter 5 The Decision Phase – To Collaborate or Not to Collaborate? That Is the Question: If you are faced with an opportunity and need a partner, whether they are known or new, this chapter will take you through a decision-making process that ensures you to create the foundations for a successful collaboration.

Chapter 6 Transition to Day-to-Day Operations – From Exploration to Operations: Having jointly researched the collaborative opportunity and agreed to go ahead with it, this chapter shows you how to turn these good intentions into the policies, procedures, and plans (business case, contract, business plan, exit plan, ERMP) necessary to begin operations.

Chapter 7 The Operations Phase – Business as Usual: This is what collaboration is all about and making sure that it is innovative, dynamic, and profitable for all concerned. In order to maximize value, we describe the essential management tasks and provide supporting checklists.

Chapter 8 Performance Measurement – If You Don't Measure It, You Can't Manage It: It is important to apply the basic principles of business metrics to manage collaborative performance. However, a performance measurement system that is optimized specifically for relationship management is needed. We describe one.

Chapter 9 Performance Management – If You've Measured It, You Can Now Manage It: We describe how to use relationship metrics to identify their strengths and weaknesses and understand how this knowledge can be used to track operational effectiveness/efficiency and the achievement of value and to stimulate innovation and continuous improvement.

Chapter 10 The Exit Phase – All's Well That Ends Well: All relationships will inevitably come to an end and planning for this throughout their life will make sure that the ending is smooth and amicable. Unfortunately, this key phase is often ignored with disastrous and expensive consequences. We provide detailed guidance on how to plan and implement this crucial phase.

Chapter 11 Review: A final word.

Checklists

You will notice that we use quite a lot of explanation in each chapter. This is because managing jointly is so different to what you are used to. We boil down the key points from the discussion to a series of 'must do' checklists that cover the main management activities.

Worksheets

At the end of every chapter we will provide a worksheet that will reinforce your understanding of relationship management and give you a practical view of how it can work by encouraging you to examine your organization's relationships. You will gain more value from this if you do it with your main partner's representatives. By the time you get to the end of this book, you will have a map of your current relationships and, thus, be able to upgrade their performance and construct a management blueprint for future use. This book will help you personally to develop rare capabilities and marketable skills as well as enhancing your company's reputation as a successful collaborator.

Terminology

Finally, in the world of collaborative relationships, terminology including job titles is not yet standardized. In order to achieve some precision of meaning, we have used terms that focus on a standard view of the function of relationship management. For this reason, we are including a glossary of terms here rather than tucked away at the end of the book.

Glossary

Assessment: is a process comprised of performance metrics (surveys) and the details of the underlying pointers (interviews) to understand the drivers of relationship success and failure.

C^3 Behavior: Co-operating, Coordinating, and Collaborating.

Collaborative Relationship: collaboration is where firms enter into buyer–seller relationships, supply chain partnerships, services marketing arrangements, or any other relationship combination. They will work together using their specialized resources innovatively to achieve aims and objectives that could not be realized on their own. Common types are:

- **Alliances:** two or more parties, often led by larger organizations, where they decide to work with others to achieve common objectives, e.g. in response to market or governmental pressures.
- **Customer–Supplier Relationships:** often found in supply chains and having evolved from transactional relationships.
- **Joint Ventures:** this is a more formal arrangement where two or more organizations create another organization to achieve their aims. The partners will have financial responsibility for the joint entity, usually through shares, and will often second staff to it.

Partnerships: where two or more organizations decide to create a formal or informal framework in which they can work together. Not to be confused with the legal definition.

Sales (Vendor–Vendor) Relationships: where two or more organizations work together to define and implement a go-to-market strategy.

Strategic Alliances: usually where the reason for the enterprise is of strategic importance to the partners, for instance, where the manufacture of a new car is crucial to the future survival of the manufacturer and the engine supplier.

Collaborative Value: the achievement of a combination of tangible and intangible products of collaboration that will not only positively impact the bottom line but also provide a continuously increasing stream of benefits:

Intangible: factors such as innovation, knowledge and skills capture, enhanced reputation and brand value, and satisfied stakeholders and relationship quality. The latter is characterized by harmony, trust, commitment, open communications, and a willingness to work together for the long term.

Tangible: increased profits, greater market share, and the fulfillment of project objectives underpinned by operational imperatives of time, cost, quality, and delivery.

Culture: 'the way things are done around here'.

Enterprise Business Case: is a joint perspective of the business opportunity.

Enterprise Business Plan: is the formal written document containing the goals of the joint business, the methods for attaining those goals, and the time frame for the achievement of the goals.

Enterprise Relationship Management (ERM): is the target operating model for coordinating all the business activities that are essential to the success of a joint/multi-party endeavor. It encapsulates the formal and informal management of those aspects of a joint enterprise that contribute to bottom-line success and is the key to successful collaboration.

Enterprise Relationship Management Plan (ERMP): is the storehouse of all the procedures and controls that make up the target operating model, and thus, it is necessary for managing the relationship. It will contain all the records of decisions, activities, and knowledge pertaining to the joint operation and will be made available to those who need to access it for both reference and updating.

Exit: is the process of ending a collaborative relationship. It must ensure that if alternative provision is required for any reason, an adequate level of support can be maintained without disruption to operations.

Exit Implementation Project: is the series of actions based on the policies laid out in the exit plan in order to terminate a business relationship and in some cases to transition to a new one.

Exit Plan: is a plan for an orderly dissolution of the relationship. This document is the strategic framework which will be used for developing a detailed project plan to initiate and implement an orderly exit.

Exit Strategy: is a high-level view of the potential circumstances that may cause the terminating of a collaborative relationship and those actions needed to achieve a satisfactory conclusion.

Framework Contracting: is where contracts that facilitate the evolution of highly co-operative, flexible relationships are focused on outcomes.

Goal/Objective: is a desired or aspirational result that an organization or organizations envision, plan, and commit to achieve.

Implementation Project: is the series of actions as defined in the business plan required to establish the joint enterprise.

IPR: The right of a person or company to exclusively use its own ideas, plans, and other intangible assets without competition, at least for a certain period of time. Examples of intellectual property include copyrights, trademarks, patents, and trade secrets. Intellectual property rights may be enforced by court through a lawsuit. The idea behind the protection of intellectual property is to encourage innovation without fear that a competitor may steal the idea and/or take credit for it.

Outsourcing: is a business model whereby an organization decides that it will no longer carry out a function in-house; instead this function will be provided by a third party.

Policy Framework: consists of the full business case, commercial agreement, business plan, exit plan, and ERMP. It will be modified and adapted in light of operational and environmental changes.

Relational Governance: is the agreed management processes found in social relationships which are used in complex collaborative situations to provide an adaptive, open, trusting environment. This is used in conjunction with framework contracting.

Relationship Analysis: is the process that reveals the relationship dynamics based upon data collected from surveys and interviews.

Relationship Manager (RM): is a senior executive who has both strategic and operational responsibility for relationship management.

Relationship Maturity Matrix: is a category of models that describes the current characteristics of a relationship across a spectrum from 'bad to excellent' according to a number of defining dimensions. Its main use is to enable partners to set improvement goals. In our experience it is of limited value because it does not give any pointers to where and what needs to be changed.

Relationship Measurement: is the process that generates comparative metrics that describe key relationship drivers and some additional characteristics based upon knowledgeable staff perceptions, usually collected through a survey.

Relationship Performance: are those aspects of a joint enterprise that contribute to the bottom line/value for money/productivity.

Target Operating Model: is a blueprint of a firm's business vision that aligns capacities and objectives in order to achieve success.

Transactional Relationship: is a relationship based on the purchase or supply of a commodity product or service which is widely available and price driven. The management of the relationship is limited to contract requirements of quality, time, and price.

Chapter 1 Worksheet

The purpose of this exercise is to draw up a simple map of your important business-to-business relationships and to highlight their effectiveness in achieving your objectives.

Activities

- List your main collaborative relationships
- Annual monetary value of each
- Date commenced
- What stage are they at: deciding, transitioning, operating, or exiting?
- Rate their strategic importance
- Who is your Relationship Manager for each?

- Who is the partner Relationship Manager for each?
- What management arrangements are in place?
- Rate the level of satisfaction for each (Highly satisfied, Satisfied, Neutral, Dissatisfied, Highly Dissatisfied)
- Significant performance comments
- Sort the list by satisfaction level within strategic importance

Questions

Does the picture you have created surprise you?

Do your management arrangements correlate with strategic importance and satisfaction?

Is your focus on the most important or the most troublesome relationship?

Are all of these relationships worth taking forward into the future, and if so, what effort and resources are you prepared to devote to this?

> We didn't walk away when we normally would have for some reason, don't know why, but it's worked out really well, they've turned out to be surprisingly good.
>
> **MD, Global Retail Fashion**

Chapter 2

Enterprise Relationship Management (ERM) – A Target Operating Model for Collaboration

What Is Enterprise Relationship Management?

Enterprise Relationship Management (ERM) is a target operating model containing all of the management processes for coordinating the business activities that are essential to the success of a joint/multi-party endeavor. It involves the formal management of those aspects of a joint enterprise that contribute to bottom-line success and is the key to successful collaboration.

Why Is Collaboration Important?

Companies freely enter into relationships in order to work together using their specialized resources innovatively to achieve aims and objectives that they could not realize on their own and at an acceptable cost. Although there are subtleties of managing relationships within different contexts, the underlying principles of collaboration remain the same.

The customer wants more than an efficient, cost-effective turn-key operation. He or she wants a successful, cutting-edge supplier to contribute to the achievement of his or her strategic objectives. The supplier wants a deeper, more valuable, long-term relationship with a successful customer. Bring

DOI: 10.4324/9781003221340-2

these two together within an effective relationship, and it is possible to open up new opportunities to share increased returns.

Broken Cycles Seventy-five percent of alliances fail to live up to expectations because they don't enter the continuous value creation cycle. See Figure 2.1. The reasons are many, but overall the lack of clear focus on what the value proposition is and how it is to be achieved leads to unreality in the way that the relationships develop. The partners will inevitably have different views, and if these are not reconciled into a joint one from the start, then the divergence of objectives will sow the seeds of relationship failure.

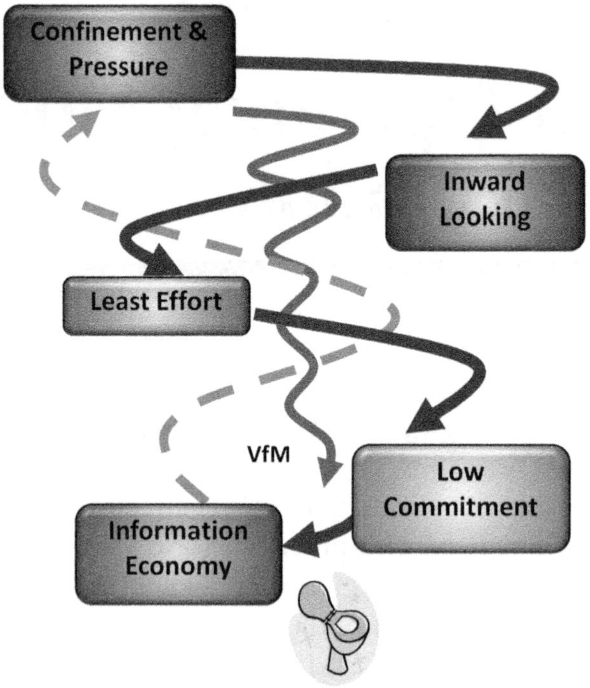

Figure 2.1 Broken value cycle.

There are a number of common situations where the value cycle is broken and value is lost:

- **Not Adapting to Change**: New requirements aren't recognized and reconciled; a gulf opens up between the partners.

  We couldn't change the relationship to access the supplier's wider expertise; therefore we terminated it and looked for a new one where we could.

  IT Outsourcer

We want to deliver a Total Facilities Management Service, but the customer just wants cleaners and security guards.

Global FM Service Provider

- **Level Blockers**: The board may well be d'accord on the benefits of collaboration, but their failure to communicate this clearly to the various levels in their organization leads to staff feeling threatened by the requirement for a more co-operative approach and resisting the development of the relationship.

  I know what they want, but it's too airy-fairy for a construction company. You try and be nice, and the customer jumps all over you.

  Chief Engineer, Global Construction Company

- **Value Confusion**: Relationship disharmony is confused with poor relationship performance, which reduces the perception of the value generated. This could result in a highly productive relationship failing.

  We hate working with them, but we can't find anybody else who does it better.

  Global Food Manufacturer

- **Feel Good/Complacency**: This is the 'party culture' where the relationship has settled into a cozy groove where 'inconveniences' such as financial governance and process improvements are ignored. Poor performance is masked by good intentions and high activity. Value capture is subordinated to an easy ride, and things start to go 'south' when the bills come in.

  We built them a terrific electricity sub-station, but unfortunately it flooded the first time it rained. They still paid us for it though, and the re-work.

  Power Sub-Contractor, National Construction Project

- **Bad Targets**: You continue to lay year-on-year price reductions on your supplier; does this incentivize him to work collaboratively? Or, if a partner maximizes his efficiency at the expense of the other, is this good teamwork?

They just dump these precision instruments on the dockside with no protective packaging because they are keen to go on leave after their voyage. We are only contracted to service the things but then get hammered by their commercial people for charging the extra to fix them because they are broken.

UK Defense Contractor

- **The Seized Cycle**: Innovation in the relationship has stopped, and the partners have become dependent on each other rather than interdependent. Due to market factors they are unable to find a replacement for each other and have to 'stay together for the sake of the kids'.

  We are more concerned with metrics than achievement. We spend more time 'Weighing the Pig' than feeding it!

  IT Vendor/Vendor Partner

- **Out of Touch**: Value capture has slowed because one of the partners has failed to stay aware of the other's increasing skills, knowledge, and capability.

  The reason why we have been struggling to launch a new product is because we are wedded to old technology. We have just found out that our partner can supply far more capable modern components!

  Electronics Designer/Manufacturer

- **Unrealized Dreams**: The partners are unable to convert the dream into reality. Because they fail to understand the full potential of the relationship, they devote insufficient effort to make it work and they are unable to capitalize on its potential and opportunities.

  We felt they weren't trying, so we went off to talk to a competitor. Unfortunately they found out.

  IT Vendor/Vendor Partner

What Collaboration Looks Like

More and more in today's competitive, globalized marketplace, the provision of services and products is a result of teamwork between several organizations. Relationships between organizations of any size are strategically important. If your supplier falls down at a crucial moment, then this can have survival implications for your company or for other members of your supply chain. The management of these strategic assets cannot be left to chance, and the same attention that you devote to finance, operations, HR, etc. must be applied to business relationships. Good relationship management has the capability to give you startlingly impressive results. The key player is the Relationship Manager (RM) in each company who is the pivotal point for all activity across the collaborative enterprise. More details will be provided in Chapter 3.

Collaboration through Process

Collaboration is a complex thing. It involves all aspects of organizational and inter-organizational activity. Often when organizations try to form alliances or partnerships and develop relationships, they move into new territory and they don't understand how the basic principles of management and organization should apply. What is needed is a management process that makes the achievement of collaborative performance more certain. See Figure 2.2. It will promote teamwork that brings together the many parts of the joint

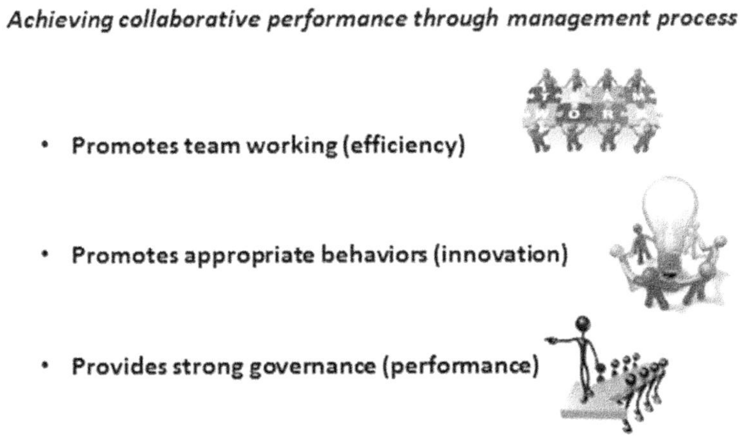

Figure 2.2 Managing for collaborative success.

enterprise into an efficient 'machine' for delivering high-value goods, services, and projects. It will also provide an environment in which innovation will thrive based upon trust, commitment, and co-operation. Last but not least, it provides a single agreed vision for the enterprise with a single view of the operations and their performance allowing clear, unambiguous control so that the result will be greater than the sum of the parts. All of this translates into increased bottom line/value for money performance.

> We have created a huge number of opportunities for dialogue including embedding staff in each other's organizations.
>
> **Engineering Director, Aircraft Maintenance**

Managing across Company Boundaries

Collaborating organizations will manage their joint enterprise using both individual and joint structures. Because of the difficulties associated with coordinating cross-boundary operations, management processes must be very clearly specified and will be dependent upon high-level agreement. Thus, the joint structure will vary depending upon the characteristics of the relationship.

The Enterprise Relationship Management operating model brings partnership management into sharp focus because it becomes proactive and accountable. It is comprised of formal organizational arrangements that ensure all activities are jointly managed at all levels. It needs to be supported by an objective performance measurement system that allows the creation of a clear, joint understanding among the partners and enables them to get things done to time, cost, and quality. See Figure 2.3. Moreover, it supports a management system that has the ability to always create and capture the maximum joint value possible within the relationship. Its success is dependent upon the commitment of dedicated resources, but it will become a valuable capability in its own right as expertise grows in the management of complex outsourcing relationships.

> We are starting to face up to performance issues at our regular meetings. In the past, mutual defensiveness got in the way of making improvements. This has been a very hard matter to tackle, but it is getting easier as we get used to working together.
>
> **Relationship Manager, Gas Turbines Manufacturer**

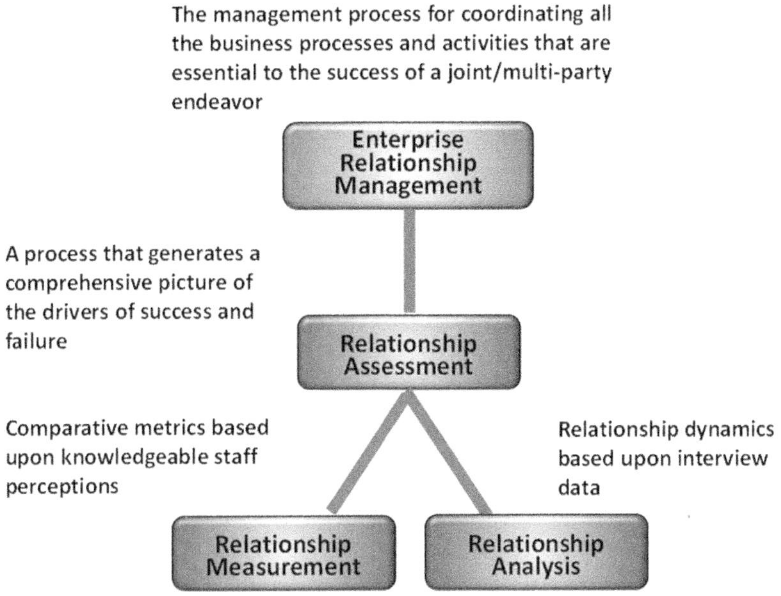

Figure 2.3 The ERM operating model.

> We were amazed by the results! We had never anticipated the additional value we could get from something right under our noses: our business relationships. Our first year cost reduction was 5% of contract value.
>
> **MD, Global Facilities Management Company**

Choosing the Right Kind of Contract

Traditional management theory on contracts states that governance has to deal with three main challenges:

- Safeguarding exclusive investments
- Measuring relationship performance
- Adapting to unforeseeable changes

Thus, companies have often relied on contract terms and conditions to manage suppliers. However, collaborative relationships are complex, strategic in nature, and their outcomes are difficult to predict. In these circumstances, writing comprehensive, enforceable contracts is extremely difficult to do.

> We have been struggling for over two years to make this partnership work. Staff at all levels are frustrated by our inability to work

together more productively. It has recently dawned upon us that this might be because we didn't re-negotiate the contract when we decided to become collaborative.

Global IT Outsourcer

Some have said that contracts are not appropriate for collaborative relationships where a 'handshake' should suffice. However, there is still a critical need for contracts to provide a legal structure to support the joint enterprise. However, they must shift from merely specifying deliverables to providing frameworks that facilitate the evolution of highly co-operative, flexible relationships that are focused on outcomes. An example in common use is the international NEC3 ECC Form of Contract used in engineering, construction, and other sectors. 'Where traditional forms of contract can promote an adversarial relationship between contractor and client, the NEC promotes a partnering, collaborative approach which has a consistent record of delivering projects on time and on budget at its core'.

The process of sitting down together to agree on value goals, operational tactics, the relationship management structure, and performance measurement will promote collaborative working. The commercial managers need to be included in the RM meeting so that they can contribute to developments and be able to maintain the currency of the contract/agreement/SLA. This continuous review process reduces the cost of commercial management and, more importantly, reduces the risk of contract breakdown and associated costly relationship disruption. Moreover, it will ensure that the relationship is flexible and constantly able to maximize its potential.

CASE STUDY: COLLABORATION BY HANDSHAKE

A pair of successful SMEs had grown their businesses together over 20 years without a formal contract in place. However, difficulties with creating and launching a new product made them realize that their processes, communications, and technical coordination had become disjointed. The lack of a clear, formal agreement for managing the relationship resulted in higher process costs, communication disconnects, loss of competitive edge, and growing hostility.

Establishing and Maintaining Joint Control

Working together over time will lead to the build-up of trust and the development of a relationship that is less dependent on the detailed terms and conditions in a contract. Positive attitudes result in:

- Flexibility – the ability to adapt in a fluid environment
- Solidarity – encouraging joint problem-solving and creating a commitment to joint action through negotiation
- Information sharing – encouraging transparency, short- and long-term planning, and joint goal setting

As the parties commit to these, the result is C^3 (Co-operating, Coordinating, and Collaborating) behavior. This fosters expectations of continuity. It encourages the partners to take a long-term view of the relationship including investments, performance, innovation, and rewards. Over time trust will develop and strengthen. Under these conditions, working together is likely to bring about increased interdependence where the parties become reliant on each other to achieve their goals. The growth of trust ensures that neither feels constrained by the loss of autonomy.

> We need to create a more open environment of trust where either party can raise an issue which gets actioned.
>
> **VP, IBM**

Relationship Performance Measurement

An important aspect of ERM is an agreed set of joint performance criteria which will feed into management processes to ensure optimum relationship efficiency and value capture. These will cover the main drivers of innovation, investment, communication, operations, and value and will ensure that the partners focus on the same factors both within their own organizations and across the partnership.

Communicate, Communicate, Communicate!

Relationships between organizations, such as partnerships, alliances, and collaboratives, are a prevalent way of working in these days of globalized, outsourced, and specialized networks of organizations. Whether in the public or private sectors they are notoriously hard to manage and the results of failures can be extremely serious.

Is it something to do with communications, sending the wrong, inaccurate, or disjointed messages? Has anyone got a view of or responsibility for the end-to-end processes? Each side thinks they have put in sufficient resources to make the enterprise work, but has anyone thought about the 'big picture'? The original vision was clear, but was it communicated to all concerned? Has it kept track of changing circumstances? Have the objectives become misaligned or conflicting? Who is responsible for encouraging and exploiting innovation, anyone? The worst thing about these uncertainties is that they tend to feed on themselves like a snake eating its own tail, and the time lag between apparent problem and management attention is so long that fire-fighting is the first rather than the last reaction of the partners. Over 50% of very promising partnerships fail in the first year. Joint communications must be used to ensure that all concerned are kept fully aware and on-side with the progress and direction of the relationship. Wherever possible these should be face-to-face or online. E-mails, texts, and other digital mechanisms are insecure, unfocused, and notorious for causing misunderstandings.

Collaborative relationships draw upon people with relevant skills and knowledge from the partnering organizations, and, therefore, there is an increased need to use communication to establish a sense of teamwork and common purpose. There is also a need to ensure that processes both within and between the partner organizations integrate. Those involved need to know not only where they fit but also how their processes interact with others both inside and outside of their company. They will also need to know what progress has been made, what current developments are taking place, and what is planned.

Personal and team accomplishments will need to be celebrated. Outside of the relationship team, both companies will need to be kept informed, and in the wider environment, individual and joint 'press releases' will need to be planned and executed, if appropriate.

The Enterprise Relationship Management Plan

The idea of a relationship management plan that helps an organization to keep track of the activities that support a business relationship is not a new one. It may go by a number of names including Alliance Project Plan, Partnering Plan, and Inter-Agency Working Protocol. We use the term Enterprise Relationship Management Plan (ERMP) specifically to indicate that it covers the entire, joint enterprise and is the key 'document' used to support our proposed management system. The ERMP should be appropriate to the size of the relationship. It must be simple, clear, and available to all those who need to access it both for reference and to record decisions, activities, achievements, and plans. It is our target operating model in use. Details of how to create and use an ERMP are elaborated in Chapter 4.

CASE STUDY: NO TOM, NO SUCCESS

A consortium was formed from a group of specialist engineering companies to provide maintenance, renewal, and upgrade of rail infrastructure. Many of the staff had originally worked within a public sector organization and had been absorbed by the new venture which was still structured in silos around the contributing companies' functions. The value of the contracted work was $20 billion over a 30-year period. This consortium was comprised of reputable companies and experienced staff, but the chance of success was significantly reduced by the failure to properly prepare for collaborative working.

Despite a high-level agreement on partners' objectives, this had not been translated into a working model for day-to-day interfaces. This led to multiple interpretations, uncertainty, and the perception that the partners' objectives were incompatible:

- Contacts with the customer were very complex with no clear mapping of managers with varying degrees of authority to many levels within each organization. This led to inconsistencies, confusion, and conflicts
- They were perceived as being as unreliable and inconsistent as a partner. It is very difficult to improve these issues and raise standards when accountabilities and interfaces are not absolutely clear
- An over bureaucratic system and instances where there was no clear decision maker led to slow and erratic problem-solving. This was sometimes compounded when, in an effort to provide resolution, formal methods were circumvented, and more confusion resulted

- They had a very clear understanding of the customer's data and performance requirements, but information was often withheld due to a fear that it might be used against them
- However, information required by other partners in the supply chain who had a critical contribution to make was poorly communicated

The consortium's failure to develop an effective collaborative operating model resulted in it going into administration 4 years later with the loss of $550 million.

The Way to Go

It is the combination of framework contracts and relational governance within the ERM target operating model that generates higher collaborative relationship performance than any governance mechanism in isolation. In particular, the effective collaborative relationship will provide greater resilience in the face of operational and commercial risks. Management of relationships is thus vital to the well-being of the organizations and their up-stream and down-stream partners.

> The new framework contract runs for 5 years at a fixed cost. This gives us guaranteed service availability and a 20% reduction in costs. The supplier is incentivized to be innovative and increase equipment reliability. In return, he can plan his work load more effectively, cut overheads, and home in on fault trends.
>
> **UK Ministry of Defense Helicopter Project Manager**

Chapter 2 Worksheet

In Chapter 1, you listed your main collaborative relationships. The purpose of this worksheet is to help you to understand why these relationships are collaborative and how they are being managed.

Activities

- Look at each of your main collaborative relationships, and list those aspects you cannot achieve without your partner
- Briefly describe any joint processes that are in place
- Briefly describe any joint management activities, both formal and informal
- Understand how you measure performance
- Draw up a plan of your lines of communication and how they work
- Review the flexibility of your contract and how it supports these activities
- Show how all these aspects are documented so that all staff and stakeholders (internal and external) are aware

Questions

Do you believe that the picture you have created is complete?

What gaps have you found?

How effective are your lines of communication and information flows to staff and stakeholders?

Overall, how is the performance of your relationships in the light of your findings?

> Quarterly, team leader-level review meetings have made a real difference.
>
> **Project Manager, Military Vehicle Manufacturer**

Chapter 3

The Relationship Manager – Clears Tall Buildings in One Bound!

The Relationship Manager (RM) will be the fount of knowledge on all aspects of ERM in their organization. Because they play a pivotal part in ensuring the smooth running and therefore optimal performance of business-to-business relationships, they need to have the authority to make decisions and issue orders. Where you find strong relationship management, there is less friction and scattering of resources, which allows the partner organizations to move forward reliably and achieve their aims.

Who Is the RM?

The RM needs to be an individual with both strategic and operational oversight who is responsible for relationship management of their organization's joint enterprises. In the smallest organizations, the role may well be an additional task of the owner or CEO. In many organizations, it may also be an additional responsibility of, for instance, a senior project manager. In very large organizations, there may be a specific department for alliance management led by a director. If no one is responsible, then relationship management is likely to be haphazard and piecemeal; it may be divided among many different departments, or it may not happen at all. Companies that

DOI: 10.4324/9781003221340-3

give proper priority to relationship management prosper. Those that don't tend to bump along with lackluster results.

> Before we had a Relationship Manager, we always had niggling problems that added to our costs. We have made considerable savings since we have someone doing the job.
>
> **Senior Manager, BT Global**

The bottom line is that the RM is key to a firm's ability to create and capture measurable collaborative value.

What Are Their Roles and Responsibilities?

RMs are responsible for developing, implementing, and then maintaining the collaborative business relationship management process throughout each organization. They become shared resources for knowledge and experience in collaborative working. Externally, they manage the relationships with networks of organizations, often involving dependencies such as key suppliers. It is likely that this web of relationship is highly complicated, as shown in Figure 3.1. It is also likely that it will not be as balanced as suggested.

The RMs will also disseminate new knowledge such as changes in market characteristics and innovative manufacturing techniques gained from

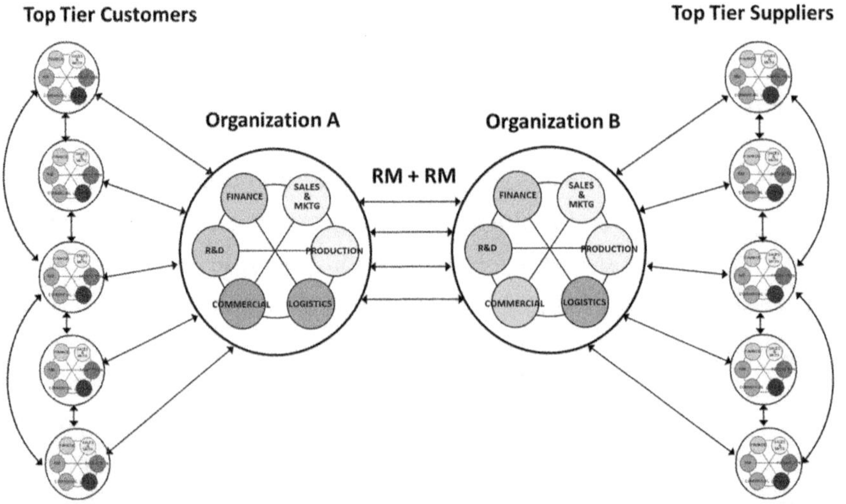

Figure 3.1 Relationship complexity.

relationship partners that can be exploited in-house. Relationship management needs to occur at these three levels:

- Having an understanding of the strategic intentions of the alliance organizations
- The in-house coordination of a firm's activities
- Coordination of activities with those of another organization(s)

Strategy Custodian The RM role has a strategic element at senior management level where decisions on partnering are made. They should be a custodian of the strategic intent of collaboration. They need to be aware of strategic decisions that could lead to selecting and defining the commercial arrangements for working with a new partner or existing partners. The RM must bring to the strategic decision-making process both policy issues for resolution and information from current partners that could allow emerging opportunities from collaboration to be exploited. The decisions from this process will be recorded in the appropriate section of the ERMP (Chapter 4).

In-House Coordination Second, the RM will be responsible for coordinating the activities of the day-to-day operations of the organization in order to ensure that they line up with the requirements for supporting external partners. This is important because different parts of the business will often have different priorities which may clash with the priorities for serving the end customer. The RM must maintain an overview of all the activities (projects, contracts, work-streams) being carried out by the organizations in order to satisfy external relationships and communicate it to all functions involved – see Figure 3.2. They must ensure that the right numbers of appropriate knowledgeable and skilled people are in place and that they are

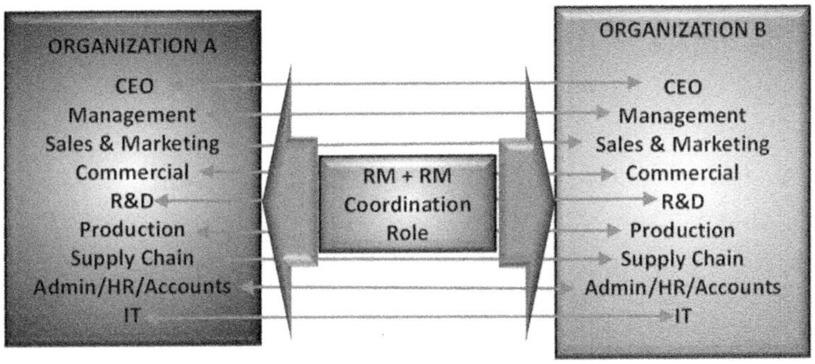

Figure 3.2 The RM coordination role.

adequately briefed. They must mediate where conflicts occur and ensure that decisions are implemented and documented in the ERMP.

Partner Coordination Third, the RM will synchronize work with their opposite number(s) in partner companies – see above. In some relationships, joint resources including people and facilities will be provided which will need to be incorporated into the joint enterprise. An appropriate level of authority is essential to enable this to happen. This is good governance which involves consistent management, cohesive policies, guidance, process oversight, and leadership. There should be regular meetings to discuss past performance, current issues, forecasts, and plans. Actions must be agreed, resourced, and monitored, and the people assigned must be held to account for their area of responsibility. Important elements from this activity will be recorded in the ERMP. The overall aim must be to maintain the relevance of the joint enterprise in line with objectives and against a background of changes in the external environment. When the relationship comes to an end, the RMs will work together to facilitate an orderly exit.

> There is an email culture within the project where anyone can talk to anyone. This brings the problem of validity and the need to balance confusion with flexibility.
>
> **Site Project Manager, Siemens Power Engineering**

Checklist of RM-Related Activities and Responsibilities

Senior Management

- Appoint experienced, knowledgeable, high-integrity people
- Give the RM department strategic status
- Keep RMs in post for reasonable periods
- Use RM advice to draft the framework contract that focuses on outputs and outcomes for the customer not on the small print

Decision Phase

- Define the objectives for the partnership including the value proposition
- Identify opportunities and target areas that the partnership could exploit
- Assess risks and operational impacts
- Quantify the benefits
- Prepare a plan to exploit the selected opportunities including costs and timescales

Transition to Operations Phase

- Identify key skills and whether training or recruitment will be needed to acquire them
- Set up and lead the joint teams
- Complete the full business case
- Complete the commercial agreement
- Complete the exit plan
- Set out the joint governance structure
- Expand the exploratory ERMP to form the target operating model (TOM)
- Adapt and modify the policy and TOM in the light of operational and environmental requirements

Operations Phase

- Maintain and manage the relationship from day to day
- Coordinate the parent company functions that service its relationships
- Jointly chair the main relationship management meetings
- Measure and monitor relationship performance
- Hold those responsible to account for their actions
- Identify and log new opportunities

Exit Phase

- Manage the exit plan

CASE STUDY: POWER PLAYS

This is an alliance of three companies with a relationship business valued at $270 million per annum. The retailer (customer) is a Big 4 UK supermarket chain where there is traditional 'beat-em-up' supplier management. They provide the channel and expect suppliers to provide its big profits. The processor (supplier) is an abattoir and innovative pork processor with specialist market knowledge. The breeder (supplier's supplier) is a highly specialized pig producer who conforms to prestigious quality standards.

The retailer rarely listens, has a siloed organization, and changes buyers to avoid building relationships. Their commercial arrangements breed fear and distrust. Their order forecasting system is inadequate and often leads to over- and under-stocking. They expect the supplier to cover the costs. Despite conforming to high-quality standards, they insist on inspecting the breeder's operations. The processor is unsettled by the lack of a long-term contract. He relies on his own order forecasting system and protects cost structure information fearing that the retailer will use it to drive down prices.

The processor's and the breeder's RMs have a long, collaborative relationship, and they work closely together to keep the retailer happy and in the dark. The quality of their products enables them to resist customer pressure and stand-up for themselves. In effect, together they are able to resist the traditional power of a very dominant partner. As a result of their efforts:

- The processor instigated regular supply chain planning meetings to improve the alliance relationships
- Planning meetings improved communications and had positive impact on overall effectiveness
- The processor was able to negotiate a long-term contract with the retailer in exchange for a cost-plus agreement
- All the parties are on better terms with each other, although the retailer is only slowly adapting as this is counter to its culture

Skills and Attributes – Communicate, Collaborate, Lead, and Serve

The RM must have leadership qualities and the ability to communicate effectively at all levels within the organization to raise awareness of its important relationships. They must have a broad view of the business and the wider industry and must have established effective networks. It is desirable that the RMs remain in post for a reasonable period so that they can accumulate experience and knowledge in this highly complex role. A company that gains the reputation for successful relationship management will attract quality partners and staff. Moreover, external stakeholders, such as investors and customers, will look favorably on the firm.

Person Specification

Thinking about the key role responsibilities of strategic collaboration, in-house coordination, partner coordination, and exit process coordination, the RM has already been noted as needing to be an exceptionally effective leader and communicator. In many roles, these skills and attributes could be seen to emphasize marketing talents. In this particularly complex role, there must also be mediation and service skills which require greater empathy, listening abilities, and awareness which ensures the managing of self and others. The following competencies and strengths are likely to be needed for the RM to be effective in developing, implementing, and maintaining the collaboration/s while working to share resources for gaining knowledge and experience and managing the network and alliances.

- **People competencies and strengths**: Able to show credible experience of engaging, developing, and leading teams; working effectively with a wide range of stakeholders; motivating teams; building loyalty and confidence within and between teams; and encouraging people to perform through co-operation
- **Cognitive competencies and strengths**: Able to show credible experience of finding practical solutions from complex and ambiguous information; learning rapidly and applying that learning; working with concepts which will assist change and continuous improvement; designing thoughtful plans, assessing priorities within the plans, then putting the plans successfully into action; and effectively managing organizational systems
- **Persuasive competencies and strengths**: Able to show credible experience of building on the ideas of others to complete tasks; assisting others in reaching decisions with a reasonable balance of risk-taking and careful behavior; and effective mediation skills ensuring parties remain engaged
- **Emotional competencies and strengths**: Able to show experience of personal resilience within difficult relationship building; creating resilience in others to reduce barriers to change and the embracing of new alliances; curiosity to find valuable knowledge and experience from internal and external partners; and high levels of self-motivation which encourages others to be motivated toward the objectives

Here is a useful 'pull-out' aide to recruiting the right Relationship Manager:

Job Specification

This role operates at board level and is responsible for the firm's ability to create and capture measurable collaborative value.

Strategic Role

- Guides strategic decisions that lead to selecting and defining the commercial arrangements for working with a new partner
- Identify emerging opportunities for collaboration and how they can be exploited
- Reporting on the performance of strategic relationships
- Ensure that policies which inform strategic decision-making are current

In-House Coordination Role

- Develop, implement, and maintain the collaborative business relationship management process throughout the organization
- Become the resource for knowledge and experience in collaborative working
- Coordinate all internal activities to ensure that they line up with the requirements for supporting external partners

Partner Coordination Role

- Synchronize work with each partner company including processes, policies, and resourcing
- Jointly manage governance and performance for each relationship

Key Accountabilities

- Responsible for managing the decision, operations, and exit planning processes
- Has supervisory responsibility for the company relationship management team
- Apply a structured framework such as an Enterprise Relationship Management Plan to relationship management activities
- Identify key stakeholders both within the partnerships and externally and gain buy-in to collaborative working
- Ensure all partners' voices are heard through the use of an effective communication plan
- Monitor relationship performance and track benefits
- Create an environment that promotes continuous improvement and innovation

Skills and Personal Qualities

- Proven leadership qualities
- The capacity and willingness to develop, organize, and manage a collaborative venture, along with any of its risks, to achieve the aims of the business
- Highly effective communication capability at all levels
- A strategic view of the business, the industry, and the wider marketplace
- Established effective networks
- Mediation and service skills requiring empathy, listening abilities, and awareness

Competencies and Strengths

- Show credible experience of engaging, developing, motivating, and leading cross-functional teams
- Work effectively with a wide range of stakeholders
- Build loyalty and confidence within and between teams and encourage people to perform through co-operation

Cognitive Ability

- Demonstrate credible experience of finding practical solutions from complex and ambiguous information
- Learn rapidly and apply that learning
- Work with concepts which will assist change and continuous improvement
- Design thoughtful plans, assess priorities within the plans, and then put the plans successfully into action
- Effectively manage organizational systems

Persuasion

- Show credible experience of building on the ideas of others to complete tasks
- Assist others in reaching decisions with a reasonable balance of risk-taking and careful behavior
- Demonstrate effective mediation skills that ensure parties remain engaged

Self-Management

- Personal resilience within difficult relationship building
- Encourage resilience in others
- Reduce barriers to change and embrace new alliances
- Curiosity to find valuable knowledge and experience from internal and external partners
- High levels of self-motivation which encourages others to strive toward objectives

Chapter 3 Worksheet

The purpose of this worksheet is to focus on the people who are involved in managing the activities that contribute to the joint businesses that you have identified in Chapters 1 and 2.

Activities

- Identify who is involved in managing your collaborative relationships
- Describe what they do
- Verify with individuals/teams if what they are doing is working and what are the problems
- Is there a single point or focus where overall responsibility for each relationship can be found?
  - If 'Yes' compare the functions being carried out for the stage that the relationship has reached using the checklist above. Identify any gaps
  - If 'No' assess the impact on the performance of the joint operation
- Understand how your organization interacts with each partner organization

Questions

Are you satisfied with the way your relationships are being managed?

When problems arise, how are they resolved?

Do you feel at both strategic and operational levels everything is under control?

Do you have people clearly identified as Relationship Managers?

If not, we recommend identifying an individual with the appropriate skills and level of authority who can be appointed as the Relationship Manager. See Job Specification above. This may not necessarily be a full-time role.

> The joint business was worth £6 million last year. We need to fund a Relationship Manager; we should have had one before now.
>
> **Program Director, Check Point**

Chapter 4

The Enterprise Relationship Management Plan (ERMP) – The 'Dashboard'

The Enterprise Relationship Management Plan (ERMP) covers the entire, joint enterprise and is used as the 'dashboard' that provides the 'instruments' and 'controls' to manage the relationship. It brings together the essential information necessary to manage and coordinate all the key activities to support our proposed management system, i.e. a collaborative target operating model.
In this chapter, we will describe each of the elements of the ERMP. We will describe what is required and why is it required. We will provide a checklist of implementation actions, and under each section we will provide guidance and examples of the contents.

What Is the ERMP?

The ERMP is a living document. It is the framework for coordinating all the business processes and activities that are essential to the success of a collaborative enterprise. It is a lens through which the joint enterprise can be brought into focus. It is the body of knowledge where all plans, activities, decisions, changes, and performances will be recorded and kept up to date. It allows progress to be monitored, risks and issues to be managed, and opportunities to be exploited. It is available to all participants and thus

DOI: 10.4324/9781003221340-4

forms a communication hub. It is also a valuable resource for inducting new team members and for briefing stakeholders.

The ERMP should be appropriate to the size of the relationship. For a simple relationship, it might be a single document with the entries listed as headings. In a more complex relationship, the ERMP would consist of a 'folder' containing documents and could be in paper/electronic/web-based format. More sophisticated systems might link diaries and databases. The format it takes doesn't matter as long as it is a comprehensive record that is used to inform decision-making, keep track of actions, and for communications. It will include static information such as the objectives, contract/agreement(s), organizational and management arrangements and policies, and contact details. It will also include dynamic information such as changes to contract schedules, regularly updated plans, performance and continuous improvement records, and minutes of management meetings including actions. At a minimum, the ERMP contains or refers to the following:

1. Organizational and Governance Structure
2. The Business Case
3. Relationship Assets Register
4. Risk Assessment
5. Commercial Agreement
6. Managing Performance
7. Continuous Improvement and Innovation
8. Knowledge Management
9. Communications
10. Personnel and Team Development
11. Exit Arrangements

The Collaborative Management Cycle

The partners will hold regular strategic reviews and operations management meetings using the ERMP as a key source of information on progress. They will also continuously review and adapt the original relationship objectives as collaboration evolves, new opportunities emerge, and in response to internal and external changes in the environment. See Figure 4.1. Many relationships do not carry out this joint activity and get into trouble because they drift out of step. They fail to achieve benefits, often suffer financial losses, and end up in court. This management review cycle is illustrated in the diagram.

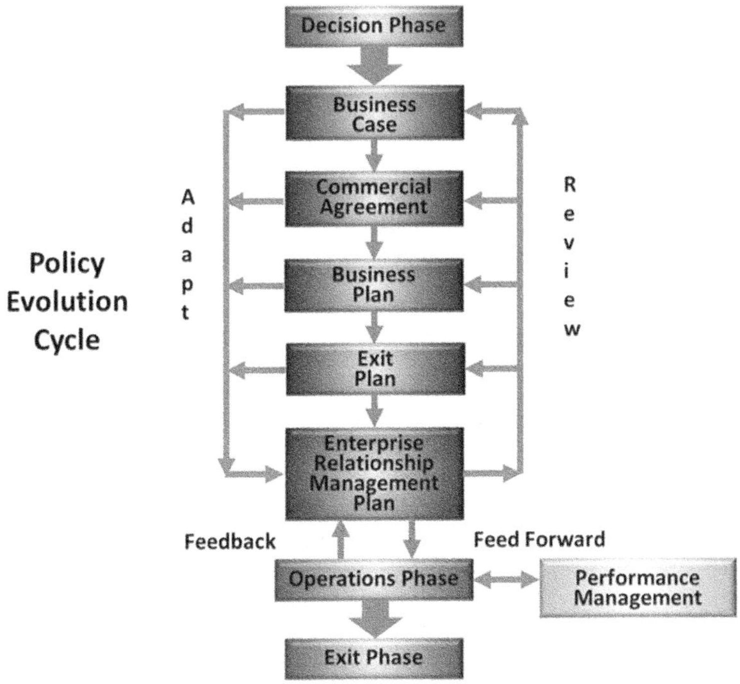

Figure 4.1 The policy evolution cycle.

Stepping through the ERMP

We now go through each of the 11 sections of the ERMP and describe what is required and why it is required. We provide guidance and examples of contents and brief checklists of important actions.

Organizational Arrangements

Partnership cohesion will be created when the organizational arrangements are recorded, maintained, and communicated to all stakeholders so that all participants are fully aware of their roles and responsibilities in the relationship and those of everyone else in the team. This section provides a map of the organizational arrangements. The details of the activities within that structure are covered in Management Activities, Continuous Improvement and Innovation, Knowledge Management, and Communications sections.

Checklist for Organizational Arrangements

- Nominate and appoint an RM from each partner organization (if not already done)

- Produce a high-level map of the joint processes (provides the rationale for the structure)
- Produce a governance structure which describes the management arrangements and terms of reference including:
  - Senior management performance reviews
  - Operational oversight meetings
- Identify the departments/stakeholders and individuals who play a key role in the enterprise
- Chart where the people involved in the relationship are located within their company's organization structure
- Record all contact details
- Record their roles and responsibilities
- Where specific joint environment knowledge is needed, it should be identified and recorded as well as the measures necessary to achieve it

Organizational and Governance Structure The structure of the joint organization must be clearly visible to all the partners in the project so that there is no misunderstanding of who does what, where they are located, and how they can be contacted. The effectiveness of the collaboration will be increased by ensuring individuals or teams take on relationship responsibilities that span all the organizations. These will cover areas such as commercial, operational processes, risk, resourcing, and performance.

High-Level Joint Process Map

For example, Company A makes products and Company B collects and delivers products to Company C who distributes them.

Organization Charts (Highlighting the Posts with Enterprise Responsibilities)

For example, Company A Org Chart → Company B Org Chart → Company C Org Chart

Governance Structure

For example, senior management committee, RM's committees, terms of reference, frequency of meetings

Contact List

Names and contact details of RMs and all other team members are recorded as per Table 4.1.

Table 4.1 Example Contact List

Name	*Company*	*Role*	*Contact Details*
e.g. *Jane Smith*	*Company A*	*Relationship Manager*	*Phone #, email*
Bob Jones	*Company B*	*Relationship Manager*	*Phone #, email*
Larry O'Dare	*Company C*	*Relationship Manager*	*Phone #, email*

Business Case

The business case is the policy document which is a statement of the overarching reason for the collaboration. It includes a detailed description of the business opportunity including the strategic alignment of the partners, the internal and external impacts, and environmental and social responsibilities. It will include the alternative options and cost–benefit analyses which led to the decision to choose this solution. Finally, it will briefly describe the proposed project to address it.

The business case is the rationale for the relationship, and the commercial agreement will be derived from it. It must be reviewed regularly to ensure that it meets the continuing needs of the partners and significant divergence is prevented. Changes will be reflected throughout this document and impact on the way the joint enterprise is managed.

Checklist for Business Case

- Place a copy of the business case and the Enterprise Business Plan in the ERMP
- Extract the Objectives and record as per Table 4.2
- Extract the Value Proposition and record as per Table 4.3
- Maintain currency as the business case changes

Objectives

This is the business rationale for the joint enterprise. It lists partner objectives and joint objectives. Each partner will have their own objectives for collaborating, and these should be made clear to all. The joint objectives will be a consolidated view of all the partners (Table 4.2).

Table 4.2 Example Objectives

Company A	*Company B*	*Company C*	*Joint*
e.g. New opportunity to satisfy	*Expand operations*	*Expand operations*	*Establish a competitive advantage*
Increase production	*Establish in new market*	*Increase utilization of existing infrastructure*	*Establish a reputation for the collaborative*

Value Proposition

Value includes revenue and new capabilities which will change over time as new opportunities arise. Where possible these should be expressed as hard targets and should be regularly reviewed (Table 4.3).

Table 4.3 Example Value Propositions

Objective	*Benefit*	*Progress*
e.g. Profit target £3.5 million in year 1	*Increased income*	*On target at Q2*
• *New production facilities in Eastern Europe* • *Factory completed in Q1* • *Production full capacity by Q2* • *Cost reduction target – 25% by Q3*	*Increased capacity at reduced cost*	• *Met* • *On target* • *On target*
Establish new transport routes	*Expanded operating experience*	• *Recruitment complete* • *Training complete* • *Operations begun*
Utilize local resources	• *Support local economy* • *Reduce carbon footprint*	• *Local contracts let* • *Monitoring begun*

Enterprise Business Plan

This lays out how the joint objectives will be achieved and likely time-scales. It supports the business case and contains a system for tracking the achievement of benefits to meet the declared objectives of the joint enterprise. It will need to be updated regularly and as circumstances change.

Relationship Asset Register

This Asset Register consists of a list of complementary resources (including financial and infrastructure) and capabilities (such as market/technical knowledge, specialized skills, and relationship management experience) that the partners draw upon to achieve their objectives. It will also include a review and matching of the cultural strengths of each partner. Culture matching is a powerful asset within enterprise relationship management.

Much of the content of the register will originate from the business case. As the enterprise progresses, new resources and capabilities will emerge, and new requirements will be identified. Also any organizational changes that impact on culture and behaviors will be recognized. These must be added to the register, which must be reviewed regularly.

A coherent, up-to-date view of resources and capabilities is essential for effective enterprise relationship management. Furthermore, the successful combination of the differing ethos and mind-set of each of the partners will enable a trusting, co-operative environment to be created which will greatly enhance the effectiveness of the collaborative enterprise.

Checklist for Relationship Asset Register

- Extract the list of Assets from the business case and record as per Table 4.4
- Complete the detail which is not available in the business case and record as per Table 4.5
- Regularly and formally review and update the register

Register of Resources

Table 4.4 Example Resource List

Description of Resources	*Provider*
e.g. New factory	*Company A*
Specialist personnel	*All*
Transport infrastructure	*Company B*
Distribution network	*Company C*

Register of Capabilities

Table 4.5 Example Register of Capabilities

Description of Capabilities	*Provider*
e.g. Product design	*Company A*
Product-specific specialist logistics knowledge	*Company B*
Local knowledge	*Company A*
Distribution planning	*Company C*

Culture Matching

Collaborative relationships are necessarily close. Are you prepared to work within the constraints of your and your partners' cultures? Simply linking or sharing business processes may not be enough to ensure success unless they can find common ground that will enable them to work together without friction. Each partner will have described their way of doing things (e.g. sharing data, transparent communications, 360-degree reviews, social activities, etc.), i.e. their organization's culture. A list of the relationship operating principles which will bring together the strengths of the respective cultures is necessary. Where possible these should be expressed as hard targets as per Table 4.6 and should be regularly reviewed.

Table 4.6 Example Cultural Strengths

Cultural Strength	*Benefit*	*Progress*
e.g. Staff suggestion and reward scheme (Company A)	*Generation of new ideas*	*Implement system by Q3*

Relationship Management Capabilities

Companies traditionally manage using standard methods such as management plans, project plans, and financial statements. Whereas these will continue to be essential in operations and financial management, they are inadequate for relationship management because they have insufficient scope to deal with the complexity of collaborative operations. Furthermore, they tend to manage relationships on a piecemeal basis, with responsibilities scattered across functions and no formal focus. Often the result is fire-fighting and poor teamwork.

Table 4.7 lists the management arrangements that are in place to make the collaborative enterprise successful based on the partners' previous experience and practice. Where possible these should be expressed as hard targets and should be regularly reviewed.

Table 4.7 Example Management Arrangements

Relationship Management Arrangements	*Progress*
e.g. ERMP system accessed through the web (Company C)	*Make available across the joint enterprise by end Q1*

Risk Assessment

Because of the close linkages between collaborating partners, any disruptions to the joint enterprise and its stakeholders will be serious. It is thus essential that the partners collectively manage risks. The temptation for partners to deal with risks individually has the potential to undermine the entire operation.

The overall risk associated with achieving the business objectives must be assessed and recorded including the continued ability of the individual partners to contribute to the collaboration. The risk appraisal must, as far as possible, take into consideration 'softer' areas such as different ways of working because these may become costly sources of friction. It is important to reconcile the partners' individual perceptions of risk so that joint mitigating plans and activities can be developed and implemented.

The Risk Register forms part of the ERMP because it is an essential element in the management and governance structure of the joint enterprise.

Checklist for Risk Assessment

- Create a joint Risk Register
- Populate initially from the business case
- Develop joint risk minimization strategies
- Regularly review and maintain

Risk Register

List the risks that could affect the joint enterprise such as environmental, legislation, operational, logistics, and management issues. The maintenance of a collaborative enterprise requires sustained management effort. This includes continuous open negotiation, problem-solving, opportunity-seeking, and teamwork. In any close relationship, problems are bound to arise and overcoming them will sometimes be difficult. A good start is to recognize these risks and plan ahead. This list is a running risk appraisal for the long-term relationship. See Table 4.8 for example entries.

Table 4.8 Example Risk Register

Risk	*Priority*	*Plan*
e.g. Product contamination	*High*	*Prevention: Rigorous 'hygiene' systems* *Containment: Identify and remove source* *Isolate and remove the contaminated batch* *Inform stakeholders*
Unexpected costs in last quarter due to product spoilage – who pays?	*Medium*	*Analyze processes and fix problems* *Decide equitable cost distribution*

Commercial Agreement

Successful relationships need the right mix of stability and flexibility. They require a base of rules and procedures that reflect obligations and expectations, yet at the same time there must be mechanisms to support change and reflect shifting internal and external conditions. The aim will be to establish strong governance procedures that ensure good communication and early warning systems. They will be based on principles that incentivize the parties to work together to reduce the probability of risk, failure, and disputes rather than dealing reactively with their consequences. The focus should be on outcomes rather than 'small print', prescriptive contract provisions.

It is an expected and natural characteristic of collaboration that it will create new opportunities. It is a key role of the commercial staff to participate in this creativity and maximizing the value that can be attained. The management of the commercial agreement, therefore, forms an essential element of the ERMP and the governance structure of the joint enterprise.

Checklist for Commercial Agreement

- Place the joint commercial agreement into the ERMP
- Extract and record the key features in the ERMP as per Table 4.9
- Utilize commercial expertise to exploit opportunities as the enterprise evolves
- Regularly review and maintain

Commercial Agreement Key Features

This includes a list of the commercial agreement(s) key features such as commitments, service level agreements, and schedules which will impact on the operation of the joint enterprise.

Table 4.9 Example Contract Key Features

Key Features
Company A will manufacture and supply products as required to satisfy orders for the joint enterprise
Company B will provide the long haul transportation network between Company A and Company C
Company C will operate the warehousing and distribution network
Terms of the contract(s) between Company C and the end customer(s)

Commercial Agreement Updates

A copy of the commercial agreement(s) will be held in the ERMP. It will be updated over the life of the joint enterprise. A record of the updates will be kept as per Table 4.10.

Table 4.10 Example Contract Updates

Date	Description of Change
e.g. Q1	*Company A has bought an additional production facility in another country requiring different logistics services from Company B and greater distribution capacity from Company C*

Managing Performance

Traditional management practices are inadequate for a joint enterprise. Instead, a formal collaborative management process must be established to oversee the implementation of the Enterprise Business Plan (copy held and maintained in the ERMP) which emerged from the joint business case and the commercial agreement. This needs to be supported by a performance measurement system that specifically covers relationship dynamics – for details, see Chapter 8. The relationship management process will be operated on two levels:

- At the senior level, strategic reviews will examine performances in the past and present, ensure that the joint objectives are still appropriate, and look ahead to how the joint business will develop in the next period.
- At the tactical level, the RMs will meet regularly to review work schedules and progress to ensure that targets are being met. Key to this is appropriate operational and relationship performance measurement that is objective and visible to all. They will manage a system for ensuring continuous improvement and innovation that handles both current processes and new business opportunities. The RMs will consider risks, resourcing, problem-solving, and setting new targets.

The ERMP will contain a transparent, comprehensive record of decisions, actions, and outcomes to meet the business plan and to ensure that innovation and new opportunities are exploited. This is the key to effective enterprise relationship management. It provides a complete overview of what it's about, how it's run, what it's achieving, and where it's going.

Checklist for Strategic Management Meetings

Senior management meets regularly, at least half-yearly:

- RMs to be in attendance
- Review performance of the past and present

- Review joint objectives
- Consider viability
- Review and update exit plan
- Consider joint enterprise development
- Provide leadership and direction

Checklist for Operational Management Meetings

Regular operational meetings jointly chaired by the RMs, at least monthly:

- Include all key personnel from across the joint enterprise
- Review performance targets in the last period and issue statistics
- Review work/orders in progress
- Review forecasted work, sales, and orders in the next period
- Review resourcing
- Facilitate the sharing of intellectual property (IP)
- Consider and solve problems
- Review and update the Enterprise Business Plan
- Review the joint Risk Register
- Manage communications
- Agree updates to the ERMP

Checklist for Operational Review Meetings

As a result of fallout from the operational meeting, the following may need to be considered by RMs separately/less frequently:

- Include all key personnel from across the joint enterprise and as necessary external experts and stakeholders
- Monitor the performance of the relationship and manage changes
- Actively seek out, initiate, and manage process improvements
- Review and update the commercial agreement
- Review value proposition and joint objectives
- Review future plans
- Review industry and technology updates
- Identify new business opportunities to refer to senior management
- Identify policy issues to refer to senior management

- Review and update exit arrangement
- Involve other supply chain partners
- Agree updates to the ERMP from these meetings

Decision Record

This is a record of reviews, decisions, and progress and is maintained within the ERMP. It provides the decision audit trail for managing the joint enterprise.

For example, minutes of meetings, action plans, and logs

Continuous Improvement and Innovation

The prime reason for a collaborative enterprise is to enable the partners to create joint value that could not be achieved working alone. It is a basic characteristic of collaborative working to create an atmosphere that stimulates innovation. This includes providing equal opportunities for contributing ideas, a reward and recognition system, and a means of tracking benefits including cost avoidance. This will give rise to proposed improvements which should be formally evaluated and, if agreed, adopted as change initiatives (policy, process, training, and audit).

Thus out of performance measurement and out of operations will emerge the need to do things differently and better. When continuous improvement and innovation become the standard method of working together, the quality of the relationship increases and the capability of the joint enterprise to achieve more value out of new opportunities is enhanced. This becomes a virtuous cycle.

Checklist for Continuous Improvement and Innovation

An improvement process will be jointly managed by the RMs and documented in the ERMP:

- Establish joint improvement process
- Create a cross functional view
- Track ideas
- Identify potential benefits

- Prioritize initiatives
- Establish joint roles and responsibilities for implementation
- Consider organizational arrangements to encourage innovation
- Update the Business Case Value Proposition as described in the Value Proposition section
- Update the Enterprise Business Plan as necessary

Innovation Register

A record of the process for considering and adopting new ideas must be maintained in the ERMP. For example entries, see Table 4.11.

Table 4.11 Example Innovation Register

Idea and Originator	*Benefits*	*Priority*	*Implementation*
e.g. Reduce standard pack by 1 inch allowing 25% more product on pallets – Outbound Logs Mgr, Company A, Q1	*25% increased transport capacity saving £2 million per annum*	*High*	*Q2 – Company A Packaging Design*

Knowledge Management

An important by-product of the joint enterprise will be new knowledge both joint and individual, such as unique processes, equipment, product designs, sales leads, R&D results, market information, and margins. There must be an atmosphere of trust and confidence so that it will not be misused. This will ensure the quality and quantity of the knowledge contributions in a self-reinforcing cycle. An agreed process must therefore be implemented to fully exploit the shared knowledge to the benefit of the joint enterprise.

The RMs will regularly review the current knowledge, including intellectual property, from within their organizations and determine what is appropriate to share with their partners. They will be aware of their organizational and joint enterprise policies on the sharing of information and must expect to act as brokers between their senior managers and the relationships they manage.

Checklist for Knowledge Management

A Knowledge Management process will be documented in the ERMP and jointly operated by the RMs:

- Senior management agree on a joint Information Sharing Policy
- A copy of this Policy must be held in the ERMP
- The Policy must be reviewed regularly
- RMs manage sharing, creation, and exploitation
- Establish and maintain a knowledge tracking system

Intellectual Property Policy Updates

A copy of the joint IP Sharing Policy will be held in the ERMP. It will be updated over the life of the joint enterprise and include any new ideas that may be generated both internally and from external sources. A record of the updates will be kept as per Tables 4.12 and 4.13.

Table 4.12 Example Intellectual Property Policy Updates

Date	*Description of Changes*
e.g. Q5	*Increase visibility of storage and handling costs – Company B and Company C*

Intellectual Property Register

This is an ongoing record of proprietary and other contributions by the partners.

Table 4.13 Example Intellectual Property Register

Creation Date	*Description & Use*	*Originator*	*Users*	*Benefits*
e.g. Q1	*Production Techniques*	*Company A*	*Company A*	*Cost control and proprietary process*
Q2	*Local Supply Chain*	*Company A*	*Company A*	*Lower costs, quality, environmental and social*
Q1	*Tailored Transport System*	*Company B*	*Company A and Company B*	*Product quality*

Communications

The RMs must establish lines of communication and ensure that information flows are effective. There needs to be a defined communication process and plan incorporating formal and informal communications such as briefings, newsletters, intranets, social media, and shared working environments which should be documented in the ERMP. It is important that joint approvals are needed for any public release.

The effectiveness of collaboration is heavily dependent on creating a sense of teamwork and common purpose and the integration of processes both within and between the partner organizations. The organizations and stakeholders will need to know what progress is being made, what current developments are taking place, and what is planned.

Checklist for Communications

Communications for the joint enterprise will be managed centrally by the RMs:

Develop a Joint Communications Plan that includes:

- Informal methods
- Formal methods
- Internal audiences
- External stakeholders
- Detailed responsibilities
- New innovations and ideas
- Ensure widespread access to the ERMP

Communications Plan

This is a description of a communications plan for the joint enterprise. See example entry in Table 4.14.

Table 4.14 Example Communication Plan

Type of Communication	*Recipient*	*Originator*	*Frequency*
e.g. Dedicated website	*All including stakeholders*	*Joint communications department*	*Daily updates*
Senior management announcements	*All including stakeholders*	*Senior management/RMs*	*Ad hoc*
Feedback system	*Managers and RMs*	*All*	*Ad hoc*

Personnel and Team Development

The partnering organizations will have their individual personnel development policies. There will be a need for a joint enterprise view in order to ensure that the appropriate skills and knowledge are available for the benefit of the alliance. This is part of creating an environment where co-operative behavior and attitudes will thrive.

The ERMP itself will provide a valuable resource for inducting new team members.

Checklist for Personnel and Team Development

Personnel resourcing for the joint enterprise will be managed centrally by the RMs:

- Create a Personnel and Team Development Policy
- Identify skills and knowledge requirements
- Create an Enterprise Training and Development Plan
- Implement Plan including the use of the ERMP as a valuable resource for inducting new team members
- Periodically review and update requirements

Skills and Knowledge Policy

This is a statement of the joint Skills and Knowledge Policy which will include a list of principles regarding personnel and team development including details of training and team-building activities. It will be updated over the life of the joint enterprise. A record of the updates will be kept as per Tables 4.15 and 4.16.

Table 4.15 Example Skills and Knowledge Principles

Date	*Description of Changes*
e.g. Q3	*Introduction of new team induction process*

Table 4.16 Example Skills and Knowledge Plan

Post	*Skills and Knowledge Required*	*Training and Development*	*Person*	*Qual*
e.g. RM	*Collaborative Leadership* *Advanced Communications* *Project Management*	*CL course* *AC course* *PM for Alliances course*	*Larry O'Dare*	*Q1* *Q3* *Q5*

Skills and Knowledge Plan

This is a continuously updated list of skills and knowledge requirements.

Exit Arrangements

At the earliest point, within the life of the joint enterprise, it will be essential to think about a number of exit scenarios. The relationship may be dissolved, and also partners may come and go during the life of the enterprise. Consideration should be given to how the knowledge captured can be rolled forward into a new phase or project. These should be encapsulated in an exit plan which will be reviewed and updated regularly.

As the relationship has been collaborative, the exit is likely to be complex and require considerable planning and effort. Significant intellectual property rights will often be involved as well as those investments that have been made and used by the partners such as skills, materials, and infrastructure. The impact on the up-stream and down-stream members of the supply chain should not be forgotten. The actions and resources that will be needed to achieve a tidy exit will be listed in the ERMP. The RMs will have an important role to play in managing the exit process because of their knowledge and operational experience of the joint enterprise.

An effective exit plan will ensure that the impact on all the organizations and their stakeholders is positive and the opportunity to do business in the future is not jeopardized. A key benefit is learning from experience and improving relationship management capability. For more details, see Chapter 10.

Checklist for Exit

The dissolution of the joint enterprise will be managed centrally by the RMs:

- List the potential conditions under which the joint enterprise will terminate including planned completion of agreement, strategic re-alignment, and relationship breakdown
- Draw up contingency project plans
- Identify exit team and resources accordingly
- Periodically review and update plan
- Implement exit plan when needed
- Record lessons learned

Exit Plan

List of actions and resources are needed to safely end the relationship. Examples might include identifying ownership of intellectual property rights and other joint/shared assets. Consider impacts on all stakeholders including customers and suppliers and how to mitigate the impact.

For example, the partners must continue normal operations until all customer contracts have been fulfilled. Appoint team and resources and implement project plan.

Lessons Learned

List of lessons learned from running and exiting this joint enterprise must be documented.

Chapter 4 Worksheet

The purpose of this worksheet is to enable you to create a relationship management information 'dashboard'. It is strongly recommended that you involve your partner organization from the outset.

Activities

- Select an appropriate collaborative relationship from your portfolio
- Invite representatives from the partner organization to a relationship-building workshop
- As an agenda for the session ensure that all attendees have a copy of Chapter 4
- Go through each section, discuss the joint management implications, and record the agreed outcomes
- You have created a collaborative target operating model. Consider how you can implement this with your partner
- Draw up an implementation plan
- Ensure that the Board and RM level staff understand the review, adapt, and operate implications

Questions

Has this exploratory process achieved a positive result?

What are the implications for the relationship? Does anything need to change?

Do you need to change the way that you manage?

Has this brought understanding of the need to invest in relationship management?

Can you take this process forward with your other important relationships?

> We need better understanding throughout the organization of the capabilities/value proposition and mutual business benefits.
>
> **Account Manager, Imtech**

Chapter 5

The Decision Phase – To Collaborate or Not to Collaborate? That Is the Question

The Decision Phase describes the process for taking the decisions that have to be made to establish the foundations for collaborative working for two or more organizations, i.e. establishing a partnering relationship and determining how it will be run. It starts by exploring through a series of steps how a business opportunity can be exploited collaboratively and concludes with the decision to go ahead or not.

First, we talk about researching a potential business opportunity and whether or not it could be achieved through a collaborative enterprise. We then consider potential partners: if it's possible to synchronize everyone's objectives, if a viable joint business case exists, and if the potential partners are willing to invest in collaborative management. The final step is the 'go or no-go' decision. At each stage, we provide checklists of important actions. This process is illustrated in Figure 5.1.

We will briefly outline the next steps to set up an implementation project in the following chapter.

DOI: 10.4324/9781003221340-5

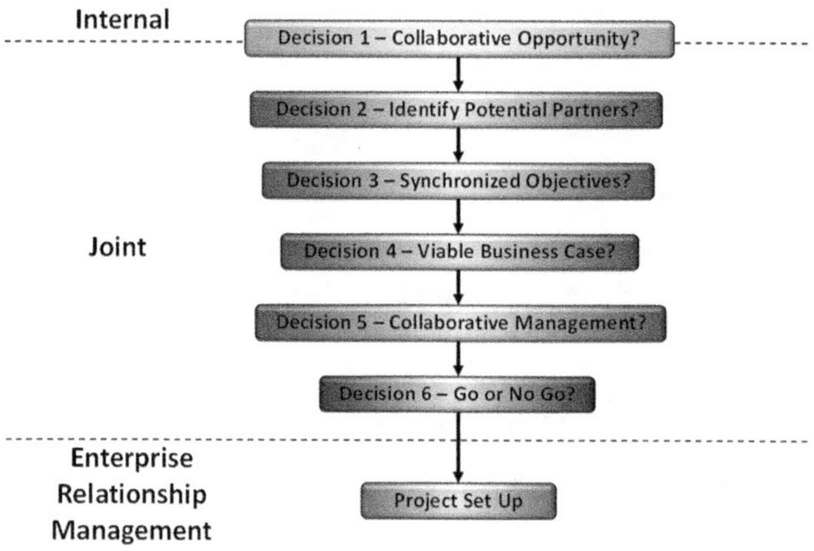

Figure 5.1 The decision process steps.

Decision One – Collaborative Opportunity?

Make or Buy? We have identified a potential business proposition that offers us significant market benefits. We consider that it fits with our strategy and objectives, but we will need additional resources to exploit it within the window of opportunity. We've considered expanding our operation, but much of the requirement is beyond our experience, and 'buying-in' the additional capability will not realize the step change in value that could be achieved. Working closely with partners appears to be the solution. But how do we do this, and what will it entail? The key question is whether we *make or buy*. It will be necessary to evaluate in detail the costs, risks, and benefits of opportunities as they arise. Only if it passes these tests will firms consider whether it is best to do it in-house or look for a partner.

Our Objectives It is essential that your organization has thought clearly about the end game, i.e. what objectives they wish to achieve. The aims of entering into a collaborative relationship including the value proposition will need to be broader than those for a transactional relationship because they will evolve and change over time. For example, possible objectives could be to access a new market, to benefit from partner know-how, to develop new products and services, and to respond more effectively to external factors. It is crucial that a company considering a collaborative relationship establishes the objective(s) clearly at the outset. It is possible that these objectives will only be broad statements of intent which will crystallize into greater detail when a partner is selected and the relationship develops. It must be recognized that these objectives will need to be reconciled with those of the selected partner(s).

> I started out delivering parcels locally. I teamed up with partners who took my business international.
>
> **Owner, SME Courier Company**

What Managers? Once it has been decided to take a closer look at the opportunity, someone needs to be appointed to lead the investigation. This person may well transition into the role of Relationship Manager if a formal decision-making process goes ahead. During the Decision Phase, the RMs should be the custodians of the strategic intent to collaborate. They need to be involved in the decisions around selecting and defining the arrangements for working with new or existing partners. They will have a key role in managing the implementation project, and they will carry this responsibility forward into the Operations Phase. A full description of the role and responsibilities is elaborated in Chapter 3.

Thinking Together Collaboration means working very closely together as a single enterprise. The partners should not just work together, but they should also think together. This has serious implications when looking at a new opportunity as it does not happen overnight. It takes time to get to know a partner sufficiently well for mutual trust to develop and to grow synergy between the organizations. If you are not prepared in this case to take this risk, then collaboration is not the option.

What Management System Is Needed? Working together collaboratively always needs more co-operative and coordinating management effort than working in an 'arms-length' relationship; therefore, partnering-specific arrangements will be needed. These components are essential:

- ERM – The formal discipline for coordinating all the business activities that are essential to the success of a joint/multi-party endeavor. See Chapter 2
- RMs – Senior executives who have both strategic and operational responsibility for relationship management. See Chapter 3
- ERMP – The joint information framework (target operating model) for coordinating all the business processes and activities that are essential to the success of a collaborative enterprise. See Chapter 4

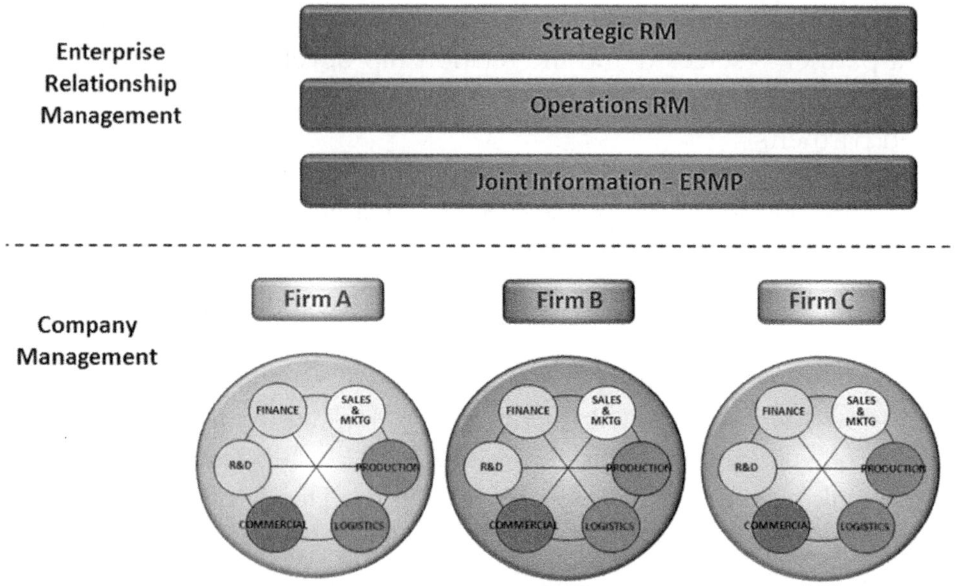

Figure 5.2 Enterprise relationship management overview.

An overview of this is shown in Figure 5.2. Depending on the size and complexity of the opportunity, if the cost of these exceeds the perceived benefits, then collaboration will not be the option.

Going Forward?

Having researched the opportunity and concluded that maximum returns can be achieved by collaborating with others and accepted that investment in a dedicated management system will be necessary, the next step is to explore the availability of suitable partners. Going forward, you will jointly explore the practical details which could become a project plan to set up a new relationship. As with any major undertaking it's important to bring together ideas, objectives, information, and plans in order to support decision-making. The aim is to create a business case for a viable joint enterprise.

Checklist for Making the Collaboration Decision

- Research the opportunity
- Consider make or buy

- Consider in-house or partner
- Understand the relationship management commitment
- Decide to collaborate

Decision Two – Identify Potential Partner?

Initial Organization Arrangements An RM must be appointed to manage the project following the criteria set out in Chapter 3. Individuals must be identified who will be involved in the activities to select a partner. If appropriate, they will need to be trained. Their roles and responsibilities must be defined bearing in mind that some of them will continue to be involved in the joint enterprise in the longer term.

Starting the Dashboard The ERMP should be started at this point to document the project so far, either to move an existing relationship to a collaborative form or to select a new partner. Having a structured approach allows for easier decision-making later on, and as the selected partner comes on board, the ERMP will change to reflect both organizations' requirements and arrangements for the joint venture. The project initiation arrangements of the RMs need to be documented in the ERMP. See Organizational Arrangements section and Exit Arrangements section in Chapter 4. It will be helpful to record the partner selection criteria in the ERMP section Relationship Asset Register of Chapter 4. These will later be changed to 'actuals' when a partner is selected. Copies of this should be produced for each potential partner and used to support the decision-making process.

So Who Do We Partner With? Collaborative relationships can be straightforward in 'customer–supplier' manufacturing, service, or project relationships. But more often they involve complex supply, co-manufacture, co-marketing, co-development, and outsourcing relationships. Usually they are of a high value and/or strategic nature. It isn't just about the rational choice around technical capability, having the resources, know-how, and reputation for reliable delivery. While these are important, of equal weight is the ability of the potential partners to work closely with us and to build up a relationship over time. It is also necessary for all partners to understand that collaborative working requires a substantial management investment. A balanced appraisal of the hard (technical) capabilities and soft (collaborative) capabilities comprise the selection decision. The RM will be responsible for carrying out the potential partner identification process.

Where Do You Find Collaborative Partners? It is likely that the potential partners will already be known because the field will not be large, and those companies capable of meeting your strategic objectives will be even fewer. They may be found from your immediate or wider network. You may not necessarily have worked very closely with them in the past, but it is likely in specialized fields that you may have had some interaction and will be aware of each other's backgrounds and capabilities. In this case, collaborative working will be easier to establish. You may also find potential partners through trade associations, then by advertising in the wider market. If they are unknown to you, then it is likely that the relationship will take longer to establish.

CASE STUDY: WE COMPETE ON THE SHELVES, NOT ON THE TRUCKS

The Relationship Managers attended a UK Institute of Grocery Distributers collaboration workshop about reducing running costs and carbon footprint by better utilization of delivery vehicles. They realized that even though they were competitors, they were in the same market, operating in a similar manner using similar equipment, procedures, and commercial arrangements. However, they had to overcome their company's cultural attitudes against working with a competitor. Their slogan was 'We compete on the shelves, not on the trucks'. Pladis sold empty space on its vehicles to Nestlé to consolidate deliveries and balance out supply and demand where there were geographical synergies. To avoid complex contractual arrangements, they agreed some basic rules of engagement and put clear boundaries in place to protect both parties' interests. They decided from the outset to limit their collaborative partnership to internal freight movements only. They put processes in place to ensure that they conformed to anti-trust legislation and used market rates to avoid accusations of undercutting third-party suppliers. Outcomes were:

- Savings of over 280,000 km of empty trailer movements, 95,000 liters of diesel, 250 tons of CO_2, and £300,000 running costs annually
- It is widely reported as being the most successful ongoing example of co-opetition in the UK logistics industry resulting in a number of prestigious awards

- Both companies acknowledge that the capability they have developed is being used with other partners
- Nestlé and Pladis are re-energizing their relationship and working together to extend the boundaries of the collaboration
- Regular joint meetings were initiated involving all the staff to seek new ideas and discuss future plans

This is a classic case that demonstrates collaborative working between competitors can access quantifiable operational, financial, and environmental benefits that were not available through regular partnerships. It also proves that it is possible to manage co-opetition successfully and remain within the regulatory anti-trust constraints. The developed capability represented a substantial competitive advantage which is utilized with other partners.

> The operation was set up with modest ambitions, and we both believe that its longevity is due to keeping it simple with straightforward governance and cost allocation mechanisms and no grand plans to redraw the complete supply chains of either company.
>
> **Logistics Director, Nestlé**

What Are You Looking For? Factors to be considered include known capabilities, reputation, market strength, and culture. Any nominations should be balanced against internal strengths and weaknesses, focusing on finding complementary skills, resources, and capabilities. A common failing with partner selection is that companies are not objective in their assessment of potential partners and what can reasonably be expected of them. Like choosing a human partner, you may want to strike up a relationship with the most successful, pretty, or athletic person in the room, but they may not share your enthusiasm or interest! You would be better off looking at your sources of value and competitive advantage elsewhere, in companies that have a similar interest and a willingness to achieve success through interdependence with you.

> Although we are comparatively small, they could supply the manufacturing capacity and we could supply the know-how.
>
> **Production Director, International Specialist Confectionery Manufacturer**

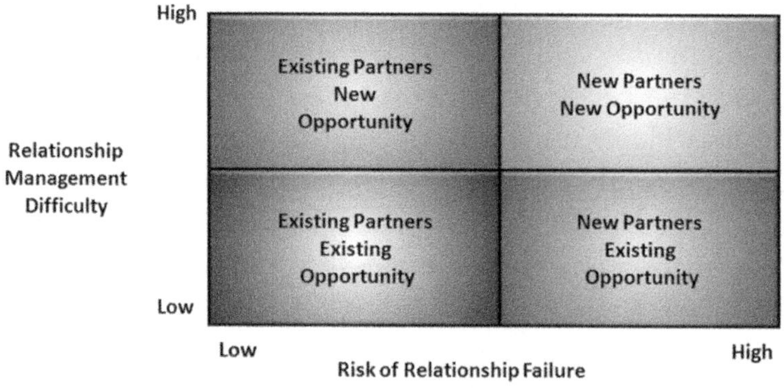

Figure 5.3 The collaboration scenarios.

The Collaboration Scenarios There are four possible scenarios where partners will need to be integrated into a collaborative team. See Figure 5.3.

1. Making an existing network more collaborative
2. Introducing existing partners to a new opportunity
3. Introducing new partners to an existing collaborative opportunity
4. Introducing new partners to a new collaborative opportunity

Developing close relationships with other organizations is going to be an exploratory process while familiarity and trust build up. The more challenging the scenario, e.g. working with new partners to address a new opportunity, more relationship management effort is required and the greater the likelihood of relationship failure. However, hopefully as the new relationship 'beds down' and becomes high performing, this risk and effort will reduce so that the collaboration becomes business as usual.

Can They Deliver What We Want? Cost control and operational reliability are critical, and the capability to deliver is a key part of partner selection. Many organizations still focus on technical compliance rather than capability, i.e. building a bridge rather than providing a means of crossing the river. Their contracts are heavy with detailed requirements that will stifle a close relationship. An understanding of the potential capability to work together to meet joint objectives is more important than hard evidence of past achievements. Collaboration is all about innovation and flexibility in both the technical and management areas and requires framework contracting to provide the right environment to achieve these. Looking at potential partners you should ask:

- Does the supplier have useful knowledge?
- Does the supplier have a leading position in its market?

- Is the supplier an innovator?
- Could the product be developed to suit anticipated future requirements?

If any of these are true, you could be looking at a future collaborative partner, and this bears on the decision to select the particular supplier.

Exclusive Investments Are there specialized human and/or physical resources that each party contributes to the relationship? For example, an information systems supplier will have to provide a customized service to the client. Similarly, the client will have to develop an understanding of the supplier's procedures, approach, and language to effectively utilize his service. Both parties may thus become 'locked in' to the relationship. This dependence on each other requires both parties to acknowledge that management effort is needed to maintain positive outcomes and to avoid the friction that can arise from feelings of claustrophobia.

Can We Work Together? Collaborative relationships are necessarily close. Simply linking business processes will not ensure success if the way business is carried out by the partner organizations are not compatible. For example, where one organization is centralized, product-focused, and bureaucratic and the other is decentralized, flexible, and market-focused, it may be difficult to find common ground that will enable both to work together without friction. Factors to be considered include known capabilities, reputation, market strength, and culture. Any nominations should be balanced against internal strengths and weaknesses, focusing on finding complementary skills, resources, and capabilities. The ability to collaborate successfully is an important capability in its own right and one of the key deciding factors in the selection process. The demonstration of past experience, a positive attitude, and an innovative approach to relationship management are desirable traits in a potential partner.

> We are trapped – we do not have manufacturing capability ourselves. However, we are not struggling to escape because they deliver despite being inflexible and quirky.
>
> **Category Manager, Global Confectionery Manufacturer**

> Although we operate within semi-autonomous business units and we are completely siloed, we somehow get on really well and the joint business has been a great success.
>
> **Food Technician, Sainsburys**

Checklist for Identifying Potential Partners

- Appoint a Relationship Manager
- Start recording dashboard information
- Shortlist likely partners considering:
  - Complementary capabilities
  - Reputation
  - Market strength
  - Culture
- Have you taken into consideration the relationship management complexity of the business opportunity?
- Do they have complementary skills, knowledge, and capacity that could make our proposition a success?
- Is their reputation for innovation, reliability, and customer focus likely to enhance ours?
- Are their customers and suppliers known to us for their reliability and collaboration?
- Is their management experience and attitude to collaboration likely to make a strong contribution to a collaborative relationship?
- Are they used to sharing confidential information in a transparent and trusting way?
- Is their structure and governance sufficiently flexible to accommodate an evolving relationship where the outcomes could be unknown?
- Can we work together?
- Record partner selection criteria in the ERMP

Our Ideal Partner? A strong indicator that you have a good partner candidate is the confidence that they will do what they say they will do, they can be trusted to share in an open and transparent way, and we will both be comfortable in a situation where we are interdependent. At this point, a clear picture should be emerging of the selection criteria and should be recorded in the embryonic ERMP section Relationship Asset Register (Chapter 4).

Decision Three – Synchronize Objectives?

Jointly Evaluating the Opportunity At this point, potential partner companies are invited to evaluate the collaborative opportunity including culture, attitude to relationship management and to risk, and decide whether it meets their strategic objectives. Our aim is to understand whether or not they can meet our partner selection criteria.

This is the exploratory step where your objectives and those of each potential partner company are matched against the proposed business opportunity. Aligned objectives are key to collaborative relationship success. The partners need to explore whether it is possible to create joint objectives. This involves achieving a deep understanding of each goal, what it means for each partner, and what it means practically in a joint enterprise.

Aligned objectives are key to collaborative relationship success. For example, one partner might want to develop new capability, whereas the other might be looking to expand into new markets. The result must be coherent joint objectives for the relationship that encompass the differing individual needs of the partners, such as:

> We are hoping to develop a new product that will satisfy a new market.
>
> **Senior Buyer, Retail Fashion Company**

They will look more closely at the tactical and strategic factors that indicate they will benefit from partnering in this collaborative enterprise. It must be recognized that to achieve a successful collaboration it may be necessary to share confidential plans and ideas with potential partners in order to select the right one. These initial negotiations will need to be covered by a non-disclosure agreement. Given the high degree of mutuality that is desired in a collaborative relationship, you are likely to have to convince a prospective partner to join forces. In practice, many of these relationships often evolve from long-term transactional dealings where trust and familiarity have already built up.

> It's not about liking each other, it's about respect for each other's capabilities.
>
> **Senior Buyer, Retail Fashion Company**

Matching Culture It's important to understand the ethos and mind-set of a potential partner and envisage whether culture matching could take place and, therefore, if collaboration would be successful. Comparing culture similarities and differences against the backdrop of the partner selection criteria, evaluate how they might work in practical terms and record in the ERMP Culture Matching section of Chapter 4.

- Are their commercial people willing to be flexible, i.e. do they use framework contracts which accept the inevitability of change, or are they tied up in detailed SLAs and penalty clauses?
- Do they expect to manage the relationship through account managers who mainly focus on the next sale?
- On the other hand, are you prepared to work within the constraints of your's and your partner's cultures?

> They are the most frustrating organization to deal with. Everything has to be referred up the line. But their customer service is second to none and benefits us enormously.
>
> **Senior Manager, Global Telecommunications Company**

Attitude to Relationship Management Organizations tend to manage relationships on a piecemeal basis with responsibilities scattered across functions and no formal focus. See Figure 5.4. It is advisable to seek partners that have a cohesive view of how to manage joint enterprises and have a track record of success as a result. This should be considered as a selection criterion in the ERMP Register of Capabilities section (Chapter 4).

Figure 5.4 Piecemeal relationship management.

Risk Considerations Risk needs to be considered on two levels. Firstly, the overall risk associated with achieving the business objective. Secondly, the risk associated with each potential partner. The evaluation process will include a joint risk appraisal.

Attitude to Risk It must be recognized that a potential partner's view of risk may not be the same as your own. For example, how much money would each organization consider to be in their comfort level for investment or loss as a result of this relationship? $1 million for one and $10,000 for the other would expose a very different attitude to risk.

Power Questions Another aspect that could be considered is relative balance of power between the potential partners and, in particular, the impact of possible changes over time on the relationship. For example, a strawberry farmer doing business with a large supermarket chain was told that his fruit was going to be subject to a half price offer and he would have to bear the cost of the promotion, pushing his income below production cost. He went out of business.

Would This Partnership Enable Us to Enhance Our Reputations? This is an important question to ask. It is possible that by combining the values of our organizations we can increase stakeholder support and boost our image in the marketplace by promoting important features such as sustainable and ethical practices.

Can We Work Together to Deliver Reliably? Our partner will need to have robust capabilities and the resources necessary to achieve the competitive advantage that we are seeking and the ability to work with us to achieve our objectives.

Can We Do Business Together? This is the last key question. Some cultural differences may be so subtle that they are not recognized as a problem until sometime into the relationship. For example, their senior management may be strongly orientated toward collaboration, but some middle management may see this as a threat to their freedom of action and obstruct new initiatives. There is never enough time and opportunity to fully examine the detailed operation and management of a potential partner, but this does not excuse a lack of a due diligence appraisal. The risk appraisal must as far as possible take into consideration 'softer' areas as potential sources of friction that may incur additional management costs. It may be that the benefits outweigh the difficulties. Document the results in the ERMP Risk Assessment section (Chapter 4).

Joint Objectives Assessment We are now in a position to put together a list of the individual objectives. These can be compared and contrasted and the potential partner's needs reconciled to create a definitive list of joint objectives. Through this evaluation it is essential to gain an understanding of the desired outcomes of the collaboration and its value. These may include potential revenue targets and other benefits such as gaining reputation, skills, and knowledge. The evaluation should include an appraisal of how the potential value can be realized, taking into consideration costs versus benefits, opportunities and risks, and resources and capabilities. The overall collaborative objectives need to be broadly stated because the detailed goals will emerge as the relationship develops. This will also allow them to be communicated to all stakeholders. A balanced appraisal of the hard and soft judgments should be recorded in the ERMP Business Case Objectives section (Chapter 4).

> I am open and above board as to where my objectives lie; they may be hard but I have no secrets.
>
> **Account Manager, Electronics Manufacturer**

Checklist for Synchronizing Objectives

- Meet likely partners and reconcile objectives
- Make choice considering:
  - Price and performance
  - Culture
  - Attitude to relationship management
  - Risks
- Are the objectives compatible with the proposition?
- Are all the partners aware of each other's objectives and comfortable with them?
- Create a list of joint objectives
- What are the critical objectives and their order of priority?
- What does it mean to achieve each objective in practice including resourcing, actions, and timescales?

Do We Have Joint Objectives? A set of synchronized objectives will provide a strong indicator to the potential partners that they will all benefit from the joint enterprise. Are they

- **Relevant**: Is the objective an acceptable starting point for the enterprise?

- **Important**: How important is each objective, and when must it be done?
- **Actionable**: What needs to be done?
- **Prioritizable**: The objectives will be sorted by relevance and importance so that the critical ones can be readily identified and fed into the outline joint business case.

If it is clear that this reconciliation cannot be achieved, then there is no point in going forward.

Decision Four – Viable Business Case?

The prospective partners have agreed that this is a potentially attractive proposition, and they have aligned and prioritized their objectives. They must now explore the business case in order to discover the practicality and extent of the value of the collaborative opportunity. The outline joint business case is the first 'hard' articulation of the joint enterprise. At this point, it can only be a high-level overview. Nevertheless, it is essential that it contains recognition that the project will be planned and run as a collaborative operation which will develop as the partners gain experience of working together to provide enhanced returns. An important input must be the prioritized objectives.

Exploring the Joint Business Case The creation of the outline joint business case is the exploratory process which will provide the partners with an understanding of the desired outcomes of the collaboration and their value. It builds upon the key objectives agreed in the previous step and recorded in the ERMP. See Chapter 4, Business Case Objectives section. It will include potential revenue targets and other benefits such as gaining skills and knowledge. This evaluation should include an appraisal of how the potential value can be realized taking into consideration costs versus benefits, opportunities and risks, and resources and capabilities.

Building the Joint Business Case This activity is designed to provide a broad indication of the business opportunity including the strategic alignment of the partners and the internal and external impacts. It should consider alternative options and cost–benefit analysis and recommend a preferred solution. It will briefly describe the proposed enterprise or business plan to address it. Finally, it will consider the individuals or team who not only will work on the business case but will crucially manage the proposed relationship throughout its life. All of this information must be added to the ERMP. See Chapter 4, Business Case section. The checklist below gives a suggested list of topics for inclusion.

Checklist for Joint Business Case

- **Business Opportunity** – A joint perspective of the proposed business opportunity
  - **Description** – What the proposed business opportunity entails
  - **Objectives and Strategic Alignment** – How the business opportunity will support the partners' strategies and strategic plans to add value
  - **Anticipated Outcomes** – How seizing the opportunity will affect the partner companies, their stakeholders, and their external environment
- **Alternative Analysis** – A summary of alternatives including the 'do nothing' option
  - **List Alternative Options** – For addressing the business opportunity
  - **Cost–Benefit Analysis** – To determine whether the project is worth pursuing
  - **Recommendations** – Summarize all of the options considered and justify the selection of the preferred solution
- **Project Overview** – A high-level description of a project or business plan to address the preferred option
  - **Project Description** – What the project will consist of, an outline of its purpose, how it will be executed, and critical information such as resourcing, timescales, and costs
  - **Project Performance** – The key resources, processes and services, who will be providing them, and how the project's performance will be measured
  - **Project Assumptions** – The preliminary assumptions for the proposed project
  - **Project Constraints and Risks** – The preliminary constraints, risks, and mitigation plans for the proposed project
  - **Major Project Milestones** – The major project milestones and their target completion dates

- **Exit Arrangements** – Partners should develop an action plan before the contract is signed. Although a standard part of the contract terms and conditions will be termination due to non-compliance, it is essential to plan for the conclusion of the relationship. The objective must be a graceful conclusion of the relationship and the potential for future collaboration. Consider:
  - Exit scenarios
  - Impact on partners' organizations and their stakeholders
  - Implementation resources needed
  - Intellectual property and assets
  - Impact on customers and the up-stream and down-stream members of the supply chain

- **Joint Business Case Management Team** – A list of the individuals who will be responsible for writing the detailed business case and implementing and maintaining collaborative relationship management for the life of the joint enterprise
  - Confirm senior representative from each partner organization (the Relationship Managers) responsible for developing, implementing, and then maintaining the collaborative relationship management process
  - Appoint team members and define their roles

Do We Have a Viable Joint Business Case? Having explored and built the draft joint business case, it will be clear whether the business opportunity will deliver the anticipated benefits and whether the proposed partners agree to go forward.

Decision Five – Collaborative Management?

Having drafted an outline joint business case and realized that the proposition is viable, the prospective partners have to explore the practical aspects of working together collaboratively. The partners should use the ERMP as a checklist of key topics to be covered in order to draft a target operating model for the management of their joint enterprise. See Chapter 4.

Exploring Collaborative Management Practicalities The use of a target operating model will provide you with a complete view of the collaborative management requirement to enable you and your partner to decide whether you can work together in it. In particular, the exercise will highlight the management resources and commitment necessary to ensure success. If this step is carried out diligently it will provide the basis for the initiation of the enterprise and a management system to support and develop the collaboration throughout its life. It will also be a central information resource about the collaborative operation and its progress available to all the partners.

Fleshing Out the TOM The prospective partners should consider at a high level each component of the ERMP in turn and insert any available information and ideas about how the joint enterprise will be managed. They will be considering a collaborative management structure in order to implement the joint business case. Their objective is to stimulate thinking about practical collaborative management, i.e. how the many strands will come together to operate as a single, joint process. At the end of this, they will have a clearer idea of the management effort and resourcing that will be needed to take the joint enterprise to its full potential. Guidance notes and example entries for each section of the ERMP can be found in Chapter 4.

Checklist for Making It Happen

- Gather the joint business case team
- Consider at a high level each component of the ERMP
- Insert any available information and ideas about how the joint enterprise will be managed
- Agree the potential management target operating model

Workable TOM? In this section, we have jointly gone through this business opportunity from a collaborative management perspective. We have explored the need to establish and operate a system of working that will extract the full potential from a joint enterprise. If we are now prepared to implement this structure, then we can make the final decision to bring the enterprise into being.

Decision Six – Go or No-Go?

A great deal of thought and effort has gone into the Decision Phase (see Figure 5.5). It has been decided that the business opportunity is best served through collaboration. Suitable partners were found with whom we could

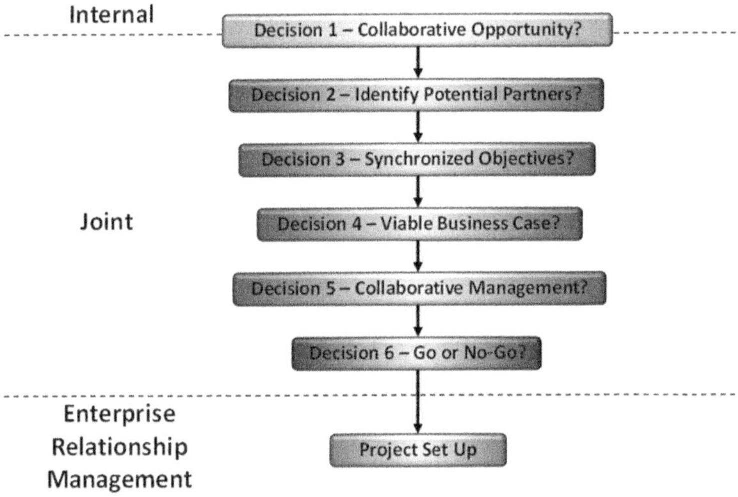

Figure 5.5 The decision process steps.

evaluate the collaborative opportunity and synchronize objectives. Together we completed a draft business case and then we explored the implications of collaborative management. We documented the key aspects of the opportunity within a draft ERMP. At this point, an informed decision, taking into consideration both hard and soft factors, can be made by the potential partners on whether or not the joint enterprise can go ahead.

Checklist for Making It Happen

- Have we agreed this is a viable collaborative opportunity?
- Have we identified potential partners?
- Could we synchronize our objectives and create an agreed set of joint objectives?
- Do we have a viable business case?
- Do we have an agreed collaborative management target operating model?
- Can we proceed with the initiation of a joint enterprise?
- Review the way in which you have handled the Decision Phase in order to learn from the experience.

Decision Phase Review – Green Light? Having carefully considered the options described in this chapter, if the answer is 'yes', then the joint project can be initiated; see details for transition to operations in Chapter 6. You will have selected a partner, but in the process you should also have made contacts with other organizations. The ERMP you raised for each should not be discarded because there is potential for future relationships with them. If the answer is 'no' as long as the partners agree, the detailed results of the decision process may be kept for use in the evaluation of future opportunities. Finally, consider the Decision Phase process to understand what went well and what didn't. Ensure that these lessons learned are reflected in your collaborative management capability.

> We have come out of the 'dark ages' and still lack a consistent approach to dealing with our customers. We are not yet prepared for collaboration.
>
> **Integration Director, Metronet Rail**

Chapter 5 Worksheet

If you are planning to select a new partner or you intend to upgrade an existing relationship to collaborative status, this worksheet is for you. You can also use it to refine your own process in the light of experiences with your existing partners. Take note that this will involve your partner organization from the outset.

Activities

- If you are about to consider a new collaborative relationship and select a partner, describe the process that you use
- Consider your current relationships and if possible describe the process that was used to initiate the collaboration
- In both cases, review your arrangements against the steps described in this chapter identifying the differences

Questions

In the course of your research studies did anybody mention possible scenarios for the ending of the relationships?

In the first situation do you plan to modify your process with what you have learned in this chapter and why?

In the second case did you get the best partner and collaboration experience as a result of your process? If 'yes' why? And if 'no' why? How would you carry out this process in the future?

> We are so frustrated by the customer we are prepared to wade in and stir them into action.
>
> **Account Manager, Augusta Westland Helicopters**

Chapter 6

Transition to Day-to-Day Operations – From Exploration to Operations

Making It Happen

Based upon the positive outcomes from the Decision Phase, we have agreed to go ahead with a joint enterprise, and this chapter describes the first implementation steps. The purpose of this chapter is to plan and set up the collaborative enterprise and its management and to smoothly transition into day-to-day operations. This is shown in Figure 6.1. It highlights the importance of continuously reviewing and adapting the policy and management operating model; as collaboration develops, new opportunities emerge in response to internal and external changes in the environment.

Full Business Case

The full business case contains the foundation for the collaborative project where the business opportunity has been defined, options have been explored, and a joint project has been outlined. The contents of the outline business case will be expanded by a joint project team from the partner companies. The full business case will contain the detailed information necessary to inform the boards in order to initiate the enterprise. It will provide the rationale which will be recorded in the ERMP. For details, see Chapter 4,

DOI: 10.4324/9781003221340-6

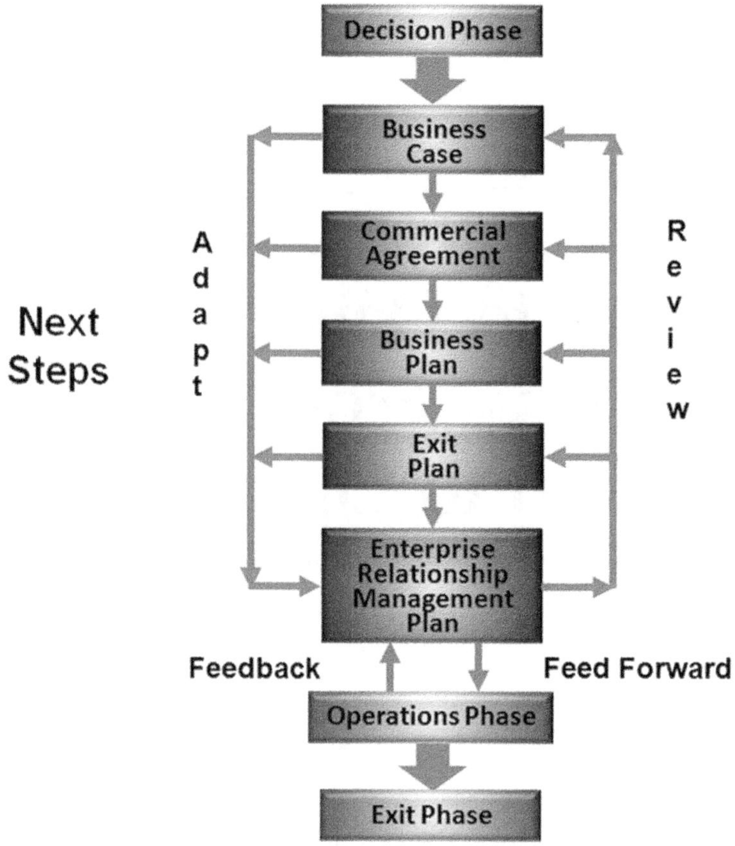

Figure 6.1 Next steps.

Business Case section. Once the parties are in agreement with the main terms of the relationship, then they will be in a position to instruct third parties, such as lawyers, suppliers, and other stakeholders.

Commercial Agreement

The agreed full business case will be used by a joint team to draft the commercial agreement along collaborative lines. For instance, the New Engineering and Construction Contract framework (NEC3) is a widely adopted, modern-day family of contracts that facilitates the implementation of sound project management principles and practices as well as defining legal relationships. The main objective is to move away from a reactive and hindsight-based decision-making and management approach to one that is foresight based, encouraging a creative environment with proactive and collaborative relationships. The agreement will need to be considered and ratified by the boards. This will provide the enterprise governance framework for the ERMP. For details, see Chapter 4, Commercial Agreement section. Because contracts and the way they are managed have such an impact on the success of the enterprise, we provide additional guidance below.

Commercial Considerations The way contracts and partnering agreements are negotiated is key to ensuring both parties get what they want and sets the tone for the relationship going forward. An adversarial agreement based on risk transference, small print, and/or terms and conditions backed up by stringent penalty clauses is unlikely to engender collaborative behaviors because it provides an environment that allows trust issues to develop. The negotiations should be pitched at the right level to focus on outputs, outcomes, the benefits, and the longer-term stability of the relationship rather than short-term opportunities.

> Now that we have a partnering arrangement around a good framework contract we just concentrate on the customer – we no longer refer to the small print.
>
> **Engineering Manager, Rolls Royce**

The aim will be to establish strong governance procedures that ensure good communication and early warning systems. It is based on principles that incentivize both parties to work together to reduce the probability of risk and failure, rather than dealing reactively with its consequence. This means thought must be given to the right allocations of risk within the relationship and about how to build and sustain trust, for example, by avoiding onerous penalties and by incorporating mechanisms for regular communications, review, and mutual performance assessments. The agreement may include joint projects or initiatives that encourage the parties to focus on continuous improvement or innovation, thereby enhancing mutual value.

> The benefits of the partnering arrangement are: we are 'future-proofed', uncertainty is removed, we can plan, and we can really focus on the customer.
>
> **MD, Food Manufacturer**

Successful relationships need the right mix of stability and flexibility. They require a base of rules and procedures that reflect obligations and expectations, yet, at the same time, there must be mechanisms to support change and reflect shifting internal and external conditions.

> I'm not sure if our success in setting up a good partnering arrangement was all luck (people) – maybe the clarity of contract aims helped.
>
> **Engineering Manager, Rolls Royce**

It is very important to get the commercial criteria right – not too tight or not too loose. Initially, the commercial criteria were included as an aid to partner selection. A copy of the final commercial agreement or a summary should be included in the ERMP. For details, see Chapter 4, Commercial Agreement section.

Enterprise Business Plan

The full business case and commercial agreement will enable a joint team to upgrade the various sections of the ERMP to form the foundation for managing the relationship. For details, see Chapter 4. For instance, in addition to resourcing and timescales, a key part of this will be the executive responsibilities, organizational arrangements, meetings and review schedules, and performance measurements which must be communicated widely. The resultant implementation plan will be used to set up the operation.

Exit Plan

It is entirely appropriate to plan for the natural conclusion of the relationship due to, for example, completion of the project/satisfaction of the objectives/changes in the marketplace. A standard part of the terms and conditions will be the need to end the relationship due to non-compliance. It will cover in principle the end of the enterprise including the distribution of assets, ownership of intellectual property rights, shutdown of activities, and the project plan needed to accomplish this. Consideration should be given to the impact on both organizations and their stakeholders and future collaboration. The impact on the up-stream and down-stream members of the supply chain should not be forgotten.

It is essential that an exit plan is written at this point. Partners should develop an exit management plan before the contract is signed. This should be held in the ERMP as detailed in Exit Arrangements section (Chapter 4) and updated regularly throughout the life of the enterprise. It will ensure a conclusion, whether planned or unplanned, is achieved that is satisfactory to partners and stakeholders.

ERMP

The ERMP (Chapter 4) will have been populated with broad concepts in the Decision Phase (Chapter 5). This has enabled the partners to explore the principles of relationship management within the context of their own business and the proposed enterprise. At this stage, the enterprise is starting to take shape, and it will be possible to add the detailed information from the full business case, the commercial agreement, and the exit plan necessary to manage the relationship. These are the policy and management components of the joint enterprise TOM.

CASE STUDY: TIGER BY THE TAIL

The customer (Global Confectionery and Snack Foods Manufacturer) initiated relationships with both suppliers (Specialist Confectionery SMEs) as part of a senior management initiative to access innovation and the rapid development of new product lines in niche markets which its own, inflexible manufacturing facilities were unable to achieve. The partner companies were small, dynamic producers of unique world-class products. Both were capable of swapping out production lines within 24 hours and developing new products rapidly. Initially, these enterprises were managed at a senior level; however, as the relationships matured they were handed over to lower-level production staff who imposed bureaucratic requirements such as daily production reports and stock returns and thus were unable to provide the individual relationship management necessary to maximize the creative returns from the suppliers.

> Their performance measures are focused at low levels rather than the overall outcome. This is the system they use with their own factories, but it does not match the service they buy from us.
>
> **MD, Specialist Confectionery Manufacturer**

In this case, it is clear that the customer had not learned how to work collaboratively. Rather than managing relationships, they employed someone to 'put out the fires':

- Senior and Line Management from both parties were unaware/amazed at the lack of knowledge of key aspects of their relationship
- Customer's management model was inappropriate
- Supplier's highly entrepreneurial, 'buccaneering' attitude was not helpful
- Communication failures on both sides resulting in unreliable logistics, penalty costs, frustrated staff, fragile co-operation, and low incentive to innovate

Informal, strategic conversations don't tend to happen so we miss opportunities to take a bigger view of the business

RM, Global Food manufacturer

Checklist for Making It Happen

Assemble the joint team led by the RMs who will brief the members on the collaborative process. They will be collectively responsible for:

- Completing the full business case
- Completing the commercial agreement
- Completing the exit plan
- Setting out the joint governance structure
- RMs expand the exploratory ERMP to form the target operating model (TOM)
- Adapt and modify the policy and TOM in the light of operational and environmental requirements
- Using the above draw up an implementation plan to initiate operations. This will become the full Enterprise Business Plan

Day-to-Day Relationship Management

Now that a target operating model has been drafted, the enterprise can move forward into establishing day-to-day management. We cover this in detail in Chapter 7 in the Operations Phase.

Chapter 6 Worksheet

The purpose of this worksheet is to help you refine your standard process for transitioning from making the collaboration decision to day-to-day operations. This will also involve your partner organization from the outset.

Activities

- So once you have decided to collaborate, now describe the process that you intend to use to move from exploration to operation
- Consider your current relationships and, if possible, describe the process that was used to implement the collaboration
- In both cases, review your arrangements against the steps described in this chapter identifying the differences

Questions

Does your process ensure that there is continuity between exploration and operations?

Are the same individuals/team involved in both?

Did the operation start smoothly, and was value generation as anticipated?

> A lot of early-design decisions and agreements were not followed by their construction personnel. Better through-project coordination on buildability is required and appropriate staff (including agents) need to be involved.
>
> **RM, Halcrow**

Chapter 7

The Operations Phase – Business as Usual

Day-to-Day Relationship Management

Having passed successfully through the Decision Phase and initiated a joint enterprise, management now moves into 'business as usual', i.e. managing the existing relationships. The secret to doing this is to utilize the combination of framework contracts and relational governance within a TOM which provides joint, formal organizational arrangements. The TOM will define the required governance structure and the broad objectives of the relationship. It becomes a living entity that adapts to the changing needs of the joint business. It promotes stability and thus supports the relationship's capacity to respond to the challenges. It provides the essential behaviors and attitudes such as openness, honesty, trust, co-operation, and long-term orientation that enable effective teamwork.

Target Operating Model – The Key to Success

The ERMP contains all the objectives, policies, procedures, and plans that make up the TOM in action and is used as the 'dashboard' that provides the 'instruments' and 'controls' to manage the relationship. For details, see Chapter 4. It enables the Relationship Managers to exercise their responsibility for operational oversight of their organizations' collaborative relationships and for working together with their opposite number(s) in the partner

DOI: 10.4324/9781003221340-7

organization(s). Together they will coordinate all the activities that are essential to the success of the joint enterprise. They focus on meeting joint objectives and both business and relational performance indicators. They instigate activities that make continuous improvement happen, and they look for new opportunities to exploit. The ERMP provides them with the necessary information to ensure that their activities are targeted on value creation. Eventually, it is used to enable a graceful disengagement within an exit plan.

Seven Steps to Success

Here are some straightforward, guiding principles that you can use to implement a successful relationship management system.

- **Business as usual** – Relationship management is an integral part of your business and should not be seen as a 'bolt on'
- **Seek authoritative advice** – Understand what you need to enhance the value of your collaborative relationships
- **Review what you do** – How can you improve the way you currently work with your partners
- **Talk to your partners** – Discuss and agree improvements to the structure for working together
- **Implement and document** – Jointly do what you agreed and keep records
- **Manage the ongoing relationship** – Stick to your TOM and keep your eye on the ball
- **Continuous improvement** – Make performance improvement inevitable by frequently reviewing with your partners how you are working together

Management Structure

The arrangements by which collaborating organizations manage their joint enterprise will involve both individual and joint structures. Because of the difficulties associated with coordinating cross-boundary operations, governance processes must be very clearly specified and will be dependent upon high-level agreements. Governance will vary depending upon the characteristics of the relationship, but there are some basic principles that should be followed. Senior managers will agree responsibilities, organizational arrangements, resourcing, funding, and performance targets both internally and jointly. The RMs will coordinate activities within their own organizations and jointly. Training may be necessary. Care should be taken to ensure that the right balance is struck between formality and informality. Too much formality will stifle innovation and involve bureaucracy. Too little control will result in costly underachievement, despite the enthusiasm this gives rise to.

These arrangements are documented in the ERMP sections, specifically 'Organizational and Governance Structure' (Chapter 4), 'Contact List' (Chapter 4), and 'Exit Arrangements' (Chapter 4). An example organization structure is shown in Figure 7.1.

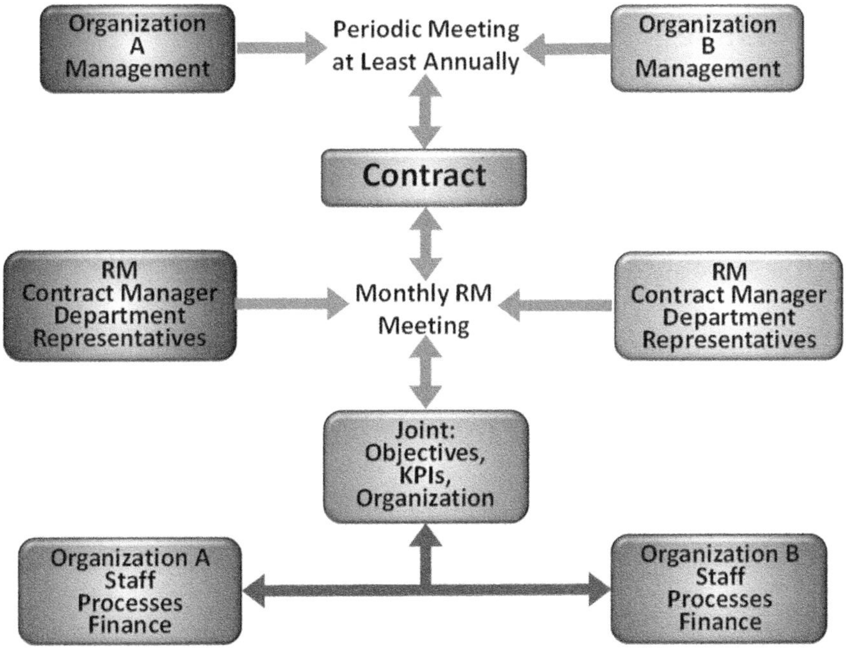

Figure 7.1 A relationship management structure.

Strategic Reviews

Regular senior management meetings should take place between the partner organizations in order to review performance in the past and present, to ensure that the joint objectives are still appropriate, and finally to look ahead to how the joint business will develop in the next period. The RMs will play a key role in these meetings.

Checklist for Strategic Management Meetings

Senior Management meets regularly, at least half-yearly:

- RMs to be in attendance
- Review past and present performance
- Review joint objectives
- Consider viability
- Review and update exit plan
- Consider joint enterprise development
- Provide leadership and direction

Tactical Management

The RM is responsible for operational oversight of their organization's collaborative relationships. They will work very closely with their opposite number(s) in the partner organization(s). As shown above, they focus on meeting joint objectives and performance indicators. The significance of the role is that it not only manages the routine but also instigates activities that make continuous improvement happen and provides the ability to exploit new opportunities that the joint business opens up. The effectiveness of the collaboration will be increased by ensuring individuals or teams take on relationship responsibilities that span both organizations. These will cover areas such as operational processes, risk, resourcing, and performance.

> As a painter and decorator, I have agreed with my partner plumber that when making good after his work I will check for leaks.
>
> **Owner, SME**

The RMs will meet regularly to review work schedules and progress to ensure that targets are being met. Key to this is appropriate operational and relationship performance measurement that is objective and visible to all. The performance measurement they use will be part of the target operating model that ensures continuous improvement and innovation and handles both current processes and new business opportunities. The RMs will consider risks, resourcing, problem-solving, and setting new targets. In accordance with the governance policy laid down in the ERMP, there will be a transparent, comprehensive record of decisions, actions, and outcomes to meet the Enterprise Business Plan and to ensure that innovation and new opportunities are exploited. All of this is effective enterprise relationship management. It provides a complete overview of what it's about, how it's run, what it's achieving, and where it's going.

At the very least, a monthly, formal operational meeting chaired jointly by the RMs would address the following:

Checklist for Operational Management Meetings

- Include all key personnel from across the joint enterprise
- Review performance targets in the last period and issue statistics
- Review work/orders in progress
- Review forecasted work, sales, and orders for the next period
- Review resourcing
- Facilitate the sharing of intellectual property (IP)
- Consider and solve problems
- Review and update the joint risk register
- Agree updates to the ERMP

> We are starting to face up to performance issues at our regular meetings. In the past, mutual defensiveness has got in the way of making improvements. This is a very hard matter to tackle.
>
> **Relationship Manager, Gas Turbine Manufacturer**

> Every month a project meeting is held where red flag issues on health and safety, costs, and quality are raised. These are minuted with actions and dates specified.
>
> **Project Manager, EDF Energy**

As a result of fall-out from the operational meeting, the following may need to be considered separately/less frequently:

Checklist for Operational Review Meetings

- Include all key personnel from across the joint enterprise and as necessary external experts and stakeholders
- Actively seek out, initiate, and manage process improvements
- Review and update the commercial agreement

- Review value proposition, joint objectives, and business plans
- Review future plans
- Review industry and technology updates
- Identify new business opportunities to refer to senior management
- Identify policy issues to refer to senior management
- Review and update exit arrangements
- Involve other supply chain partners
- Agree updates to the ERMP from these meetings

> Regular meetings to discuss lead-times, deliveries, and technical problems are essential enablers to building trust.
>
> **Design Engineer, BAE Systems**

Commercial Considerations

Some relationships are based on informal 'handshake' agreements. Others will be governed by formal arrangements such as written contracts. Whichever is in place the TOM creates a structure within which collaborating partners can develop workable commercial arrangements, including objective performance targets, open book transparency, and incentives. Just reviewing the contract at set intervals is unlikely to enhance the performance of a collaborative relationship which is subject to change and flexibility. The commercial managers need to be included in the RM meetings, so that they will contribute to developments and be able to maintain the currency of the contract/agreement/SLA. This continuous review process reduces the cost of commercial management and, more importantly, reduces the risk of contract breakdown and associated costly relationship disruption. The updates should be reflected in the ERMP.

CASE STUDY: JUST ANOTHER PRODUCTION LINE?

A global food manufacturer outsourced some seasonal product lines to a small, world-class, niche producer. The supplier was expected to fill retailers' shelves with high-quality, low-cost items. However, the customer insisted on controlling raw material sourcing and demanding weekly production statistics as if it was one of its own factories. The result was severe friction between the partners and disagreement over the performance of third-party logistics providers.

Increasing Value through Positive Action

Out of performance measurement and out of operations will emerge the need to do things differently and better. See Figure 7.2. A basic part of collaborative working is creating an atmosphere that stimulates innovation. This will give rise to proposed improvements which should be formally evaluated and, if agreed, adopted as change initiatives. One approach might be to have a joint improvement process which would:

- Have a cross-functional view
- Track ideas
- Identify potential benefits
- Prioritize initiatives
- Establish joint roles and responsibilities for implementation

> They are a small innovative company who can turn out two or three new products per year. We can only do something like one product every three years.
>
> **Senior Operations Manager, Global Confectionery Manufacturer**

Throughout, benefits realization must be tracked, using, for example, a simple spreadsheet, in order that the impetus of the collaborative enterprise is sustained. Maintaining stakeholder interest, staff enthusiasm; continuing

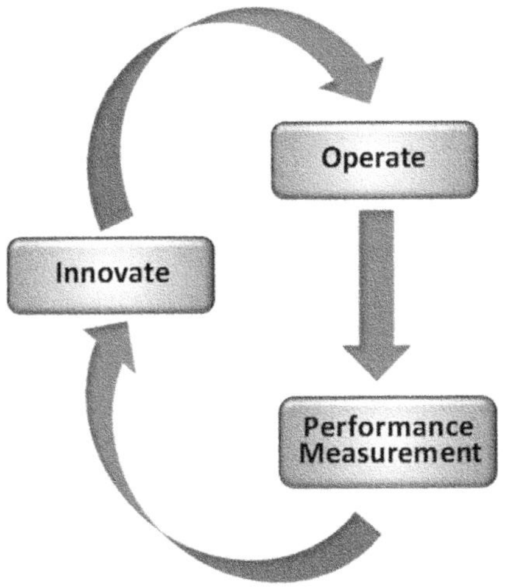

Figure 7.2 The continuous improvement cycle.

innovation, process improvements, and increased productivity are as much part and parcel of forward momentum as the achievement of the main goals.

CASE STUDY: STAFF REWARDS PAY DIVIDENDS

Her Majesty's Naval Base, Devonport. For one of our projects we publicize our achievements via joint newsletters both within UK Ministry of Defense and within the industry Partners (Babcock/Rolls Royce). Plus if someone comes up with a good idea that saves time/moneys, etc., then they are rewarded with a shopping voucher and at the very least receive a letter of thanks.

> Reward and recognition are key. We add the 'thank you element' and celebrate success together.
>
> **Partnering Manager, UK Ministry of Defense**

It is crucial that the achievement of benefits is fed back into the Enterprise Business Plan so that it can expand and develop with the relationship and its prospects. This process will be jointly managed by the RMs and documented in the ERMP.

Knowledge Management

When two or more partners collaborate, they will contribute specialized knowledge to the enterprise. Examples are product designs, sales leads, R&D results, market research, and margins. If there is confidence that this knowledge will not be misused (used without the permission of the owner), then the quality and quantity of that contribution will be greater. In these circumstances, an important product of the alliance will be new knowledge, both joint and individual, which can be exploited profitably in a self-reinforcing cycle.

The RMs will need to be aware of their organizational policies on the sharing of IP and must expect to act as brokers between their senior managers and the relationships they manage. A flexible approach is needed because the generation of knowledge and the need to exploit it, both internally and jointly, will be an ongoing feature of collaboration. There needs to be a regularly reviewed, jointly agreed policy statement in the ERMP that outlines how knowledge will be managed. Failure to manage this properly often leads to disputes and costly court cases.

Communications

Collaborative relationships draw upon people with relevant skills and knowledge from the partnering organizations, and, therefore, there is an increased need to use communication to establish a sense of teamwork and common purpose. There is also a need to ensure that processes both within and between the partner organizations integrate. Those involved need to know not only where they fit but also how their processes interact with others both inside and outside of their company. They will also need to know what progress has been made, what current developments are taking place, and what is planned.

Personal and team accomplishments will need to be celebrated. Outside of the relationship team both companies will need to be kept informed and, in the wider environment, individual and joint 'press releases' will need to be planned and executed, if appropriate. The RMs must establish the lines of communication and ensure that information flows are effective. There needs to be a defined communication process and plan, incorporating formal and informal communications, such as briefings, newsletters, intranets, social media, and shared working environments, which should be documented in the ERMP.

> The fact that we **do not** have an effective performance measurement system while our partner believes we have is a real failure to communicate.
>
> **Maintenance Engineer, UK MoD**

A Different Culture?

Collaborative relationships are inevitably deep and long-lasting and will evolve over time. Although the contract will provide the foundation for the relationship, continuous discussion and negotiation at all levels will allow innovative ideas to be captured, included in the ERMP and implemented.

Managing the joint enterprise as a team is the most effective way of getting the balance between formal and informal processes right. This is the ability to work together across organization boundaries to create the value that the individual partners could not create separately. Therefore, a different style is needed to generate the best outcomes.

Some relationships have a formal behavior charter launched with 'a fanfare of trumpets' and reinforced by regular team-building events. For others, it will just be business as usual. However it is done, the premium is on participative leadership rather than the traditional, directive style of management. It will be necessary to create a working environment where all feel that they can contribute freely; where they can trust their colleagues and their organization to go the extra mile; and where continuity, predictability, and long-term, joint gains are valued. It should be noted that trust between people and organizations will only build up over time as a result of an accumulation of small teamwork successes.

> Culture may be a hindrance in building a strong relationship. For example, while pubs and drinks are a great means of socializing in the UK, that's not the same in other countries; therefore, there is a need to actively create a forum or a social event that enables mutual participation to help seal the relationship.
>
> **Senior Systems Architect, Wipro**

Business Continuity Planning

Because of the close linkages between collaborating partners, supply chain and other disruption effects on the joint enterprise are likely to be more serious. All concerned should be alert to environmental, industry, market, political, and any other factor that might have a positive or negative impact on the operation. Analysis of potential risks and maintenance of contingency plans need to extend beyond the relationship to encompass the important members of the supply chain. Where such disruptive event could provide opportunities, these must also be considered. These should be documented in the ERMP.

> Frequent contacts, even as often as daily, build confidence, reduce risks of misunderstandings, and keep the team focused.
>
> **Team Leader, Aerospace Company**

CASE STUDY: JUST THROW IT OVER THE FENCE!

The boards decided that there was considerable value in adopting a collaborative relationship. In particular, tapping into the wider capabilities of the partners would enable more innovative ways of working, leading to lower costs and greater efficiency.

At this point, they assumed their senior management would define what was meant by working collaboratively and would identify the commercial and operational changes that were needed to secure the desired value. Unfortunately, this expectation was not fulfilled.

Senior management met with the commercial managers and the sales people. Not realizing that they needed to explore a completely new way of working with their opposite numbers, they hammered out a revised contract. This looked very much like the old one, including, for example, a demand for a year-on-year 5% price reduction, but incorporating the increased requirements that they thought the new, closer relationship should deliver and with additional governance clauses needed to manage it. Because their focus was on the bottom line, there was minimal or reduced investment. They packaged it as their new Vendor Management Program.

The new contract then hit the operational levels like a bolt from the blue. It contained expectations that they were either unable to satisfy or required considerable changes to implement and, often, with cost penalties. There was no plan for the re-alignment of the organizations to make possible a new way of joint working. As a result, staff increased their work rate in fruitless pursuit of the new targets. Frustration and friction levels inevitably rose and adversely affected performance. Eventually, penalty clauses were threatened, and the commercial managers got involved.

Increased innovation was a significant expectation of this contract, although it did not specify how it was to be incentivized or rewarded. Furthermore, new communication channels to discuss innovative value creation and capture were not established. At the operations level, frenzied activity and performance issues left very little scope or appetite for developing and implementing new ideas.

After a period of time it was evident that the collaboration was not delivering the value envisaged, and the boards began to question the original concept. The fact that the collaboration initiative was not implemented was not recognized as the source of failure. This top-level doubt started to permeate into the alliance organizations.

Chapter 7 Worksheet

The purpose of this worksheet is to consider how to ensure that your relationship continues to thrive and innovate during day-to-day operations.

Activities

- Review the management arrangements with your partner of the end-to-end operations processes. Consider, for example:
  - Overall performance (efficiency, effectiveness, and cost)
  - Problem-solving, fire-fighting
  - New ideas, opportunities, and continuous improvement
  - Value realization
  - Things going well, things not going well
- Draw up a remedial action list

Questions

Is senior management (both partners) fully aware of the performance of the operation? Is it still viable? Does it require further investment? Could it be expanded further?

How effective is the current relationship management? Is it sufficiently resourced? Is it formally structured and recognized?

> Unfortunately, the coherence of the program lost its way over time. It lost touch with reality and became a monthly fire-fighting exercise.
>
> **Relationship Manager, Halcrow**

Chapter 8

Performance Measurement – If You Don't Measure It, You Can't Manage It

The purpose of this chapter is to demonstrate the importance of applying the basic principles of business metrics to relationship assessment and how this is used to manage collaborative performance. It provides a step-by-step guide to how the strengths and weaknesses of relationships can be measured and understood.

Why Measure?

What is needed in these circumstances is a management process that makes the achievement of collaborative performance, individual and joint goals more certain. It will:

- Establish position (Benchmark, understand why; KPIs – indicators not measures)
- Communicate direction (To turn or not to turn)
- Influence behavior (Have you created a new problem?)
- Stimulate action (Take the 'right' action)
- Facilitate learning (Continuous improvement – not a 'one shot wonder')

DOI: 10.4324/9781003221340-8

It will necessarily include an assessment process consisting of performance metrics and the details of the underlying pointers to the drivers of success and failure. This will provide a strong base for joint governance. It will promote teamwork that brings together the many parts into the joint enterprise, creating an efficient 'machine' for delivering high-value goods, services, and projects. It will also provide an environment in which innovation will thrive based upon trust, commitment, and co-operative behavior. Last but not least, it will provide a single, agreed vision for the enterprise with a single view of the operations and their performance allowing clear, unambiguous control, so that the result will be greater than the sum of the parts. All of this translates into increased bottom line/value for money performance.

Principles of Collaborative Performance Management In a joint operation or enterprise, it is only logical that all sides will have agreed common performance targets as well as the coherence of their individual goals. If the collaboration is a network or consortium, performance management must involve all the partners. It must therefore be enterprise-wide and become an essential part of collaboration management. The only way to achieve this is to use an approach that provides objective performance metrics and an understanding of the detailed dynamics at both the individual and joint levels.

> We are doing a lot to educate the customer into involving the end customer in the setting of performance targets.
>
> **Project Manager, Augusta Westland Helicopters**

Why Collaboration Fails

The Same Old Story Organizations with the best intentions will partner with others in order to seize opportunities that they couldn't on their own. They set up operations on the familiar lines they have always used successfully in their own organizations. They concentrate on their own part of the enterprise and expect the others to fit in with them. As a result, gaps develop into cracks, objectives are not met, and because they do not understand why it's falling apart, tempers fray and divorce looms large. Collaborative relationships don't just happen. They need planning and discipline just like any other aspect of business. How can you expect to capture value if you don't? It is essential that managers maintain visibility and control to keep the momentum going.

It's All Too Difficult In business, we measure performance in all its forms. The tools used are sophisticated and people who study for MBAs are required to show an understanding of most of them. Why then do we not manage and measure relationship performance? The usual answer is: 'It's too difficult'. Instead, we rely on tracking operational and financial KPIs even though these are invariably in arrears and only from our own perspective. We have to trust our partner's figures and guess what is happening across our interfaces. A common problem is thus one partner may consider time, cost, and quality, whereas the other will track sales and revenues. The end result is that both use an incomplete picture to manage joint performance.

> Everyone is trying to do their best and the spirit is excellent, but there are no formal performance measures.
>
> **Site Manager, Amec**

Rear View Mirror Management Although organizations tie up huge capital and human investments in their strategic alliances, joint ventures, and partnerships, they rely heavily on backward-looking, imperfect performance measures. They then wonder how issues that have boiled away unseen such as complacency, distrust, quality failures, opportunism, late deliveries, cost over-runs, and communication gaps suddenly raise their heads as major issues. Panicked, often the first recourse is to the contract and the penalty clauses.

> Performance measurement is used but standards are never raised. Why invest more time and money for no benefit?
>
> **Maintenance Manager, Shipping**

Key Performance Management Issues Can you answer these simple questions?

- How many business relationships do you have?
- Why are they important?
- Which ones are doing well and why?
- Which are not doing well and why?
- Is there a performance gap, and what opportunity are we missing?
- What objective measurements can you use?
- How do you identify targets for continuous improvement?

- How do you do all this jointly with your partners?
- How do you do it with minimal effort and maximum effectiveness?

If you were the CEO of a major company looking to take over another or if you came into the top post of a focal organization in a supply chain or project management network, aren't these the minimum due diligence questions you would need to answer? Having seen the way many blue chip organizations operate, we can tell you that not only do they not ask them, but also they don't even know they need to ask them. This seems odd, given the so-called scientific management we declare we practice today.

We propose the answers to these questions are essential if your organization is involved in collaborative working with others. What's more, we challenge you to get the best out of your strategically critical relationships if you do not have a grip on joint enterprise performance.

Relationship Performance in Practice

The System Our research into business relationship performance has revealed the fundamental measures that focus on the underlying drivers of inter-organizational dynamics. They provide the pointers to the individual and collective, tactical and strategic activities that affect joint performance. We have developed a performance measurement system which is optimized specifically for relationship management. It is shown as an outline in Figure 8.1.

The following key performance drivers are used at the heart of this system to measure and understand the joint enterprise:

- **Innovation** – creativity
- **Investment** – alignment of objectives, resources, infrastructure, and management effort
- **Communication** – open dialogue
- **Operations** – quality service and product delivery
- **Value** – commitment to the realization of joint benefits

Relationship-specific questions are asked around each of the above drivers in a survey of knowledgeable personnel. In addition, this system provides a view of the softer aspects of a relationship such as *trust*, *commitment*, and *long-term orientation* that have a motivational impact on operational

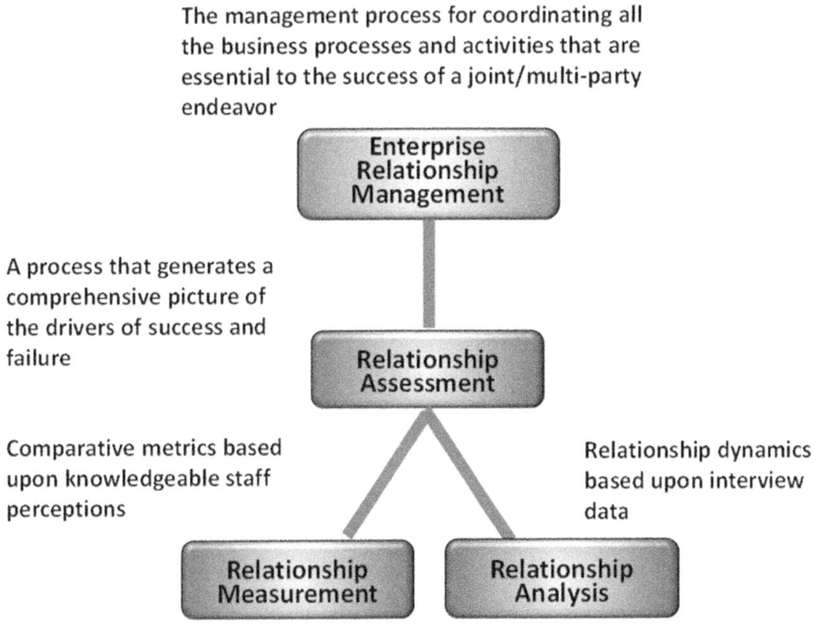

Figure 8.1 A performance measurement system.

performance. We will describe how these performance drivers are used as the basis of a total collaboration assessment system and show how they are used to manage performance in the main situations that collaborative businesses will have to deal with. First, from a practical point of view, you need to understand how performance measurement underpins enterprise relationship management and is an integral part of the TOM.

Step-by-Step Guide to Performance Measurement

The Essential Requirements In a joint operation or an enterprise, it is only logical that all sides will have agreed their common performance targets as well as the coherence of their individual goals. Performance assessment must therefore be enterprise-wide and become the essential part of joint collaborative management.

If the collaboration is a network or consortium, the assessment must involve all the partners. Collaboration assessment needs to fit in with other business measurements that trace time, cost, quality, and key events across the boundaries of the partner organizations. However, traditional performance measures are not capable of making sense of the complexity of a collaborative relationship. Relationship assessment is different. It needs to support all the business-to-business relationship functions right across the joint enterprise whether they are carried out individually or collectively. Furthermore, it should contain both metrics and relational analysis based on the views of success from the knowledgeable people who answered the survey. The result should highlight progress against goals and opportunities to reap more value from the joint enterprise. It should:

- Ideally be independently operated and interpreted by a trusted third party
- Guarantee individuals and organizations anonymity, to protect personal and commercial sensitivities and to encourage openness
- Be generic to inter-organizational relationship dynamics – not sector or site specific
- Be capable of providing a comprehensive 360-degree view of relationships including customer/supplier, networks, alliances, and consortia
- Provide a comprehensive view based upon knowledgeable team members rather than a statistical sample
- Utilize both interview and survey data to provide a rich picture
- Be quick and easy for staff to complete
- Provide outputs that will be readily usable to support managers

It was little pain for big gain.

Relationship Manager, EDF Energy

What Must Performance Measurement Do? Collaboration can have many forms and be at many different stages of development; therefore, it is necessary that assessment systems should be capable of forming an effective part of performance management in any scenario. The range of assessment situations can be:

- **'Quick look'** – A rapid gauge of performance for an initial assessment and regular progress monitoring
- **'Strategic overview'** – To discover where good and poor practices are situated in a portfolio of key relationships and allow them to be managed
- **'In-depth diagnosis'** – To measure the key relationship performance dynamics and provide sufficiently detailed analysis and recommendations to allow a change program to be initiated and sustained

- **'Partnership potential'** – To improve internal partnering capability and to help organizations choose the right partner
- **'Multi-party teamwork'** – To measure the performance and diagnose the issues within an alliance or consortium

We will discuss these in greater detail with case study examples.

What Does the Process Look Like? It should be easy to carry out, not disruptive to the participating organizations; the results should be readily understood by staff at all levels; and it should have the buy-in and confidence of all, especially senior management. In Figure 8.2, we show the three simple components of a joint enterprise relationship appraisal.

Survey:

- On-line questionnaires check Key Performance Measures
- Knowledgeable people from both companies participate

Interview:

- No longer than 1 hour, usually by phone
- Small number of knowledgeable personnel in each organization
- Capture 'headline' views

Workshop/Report:

- Headline 'traffic lights' identify 'what's going on'
- Interview quotations illustrate graphics - 'why it's happening'
- Comprehensive conclusions and recommendations
- Action Plan

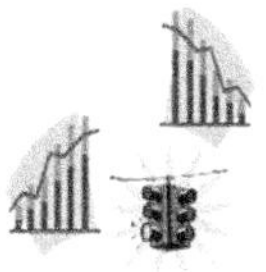

Figure 8.2 An assessment process.

Getting the Setting Right The setting for the assessment exercise is a crucial part of its success. Given the sensitivities involved, like marriage guidance counseling, the process must ideally be administered by an impartial, trusted, discrete third party. The supplier is not going to 'open-up' if the process is being run by the customer, or vice versa. The third party must be the soul of discretion because the detail that will emerge from the assessment will be commercially sensitive so a guarantee of complete confidentiality is given. The sponsors need to be the participant organizations' Relationship Managers; senior people who will represent their organizations throughout the process and will be responsible for leading the emergent change program. The third-party measurement team will work for them because it is important that the personnel see the process as being an internal initiative rather than one that is 'being done to them by outsiders'. After all, 'what do outsiders really know about the way we do things?'

Engaging with the Partners Finally, having got the undivided attention and support of the staff in the participating organizations to the assessment process, it is very important that once the results have been analyzed and presented to the sponsors, comprehensive feedback is arranged. In our experience, these staff will also be extremely keen to become involved in the resultant change program and often make excellent change agents.

Step 1 – The Survey

Who to Survey? Most managers suffer from being bombarded by questionnaires. However, this survey should be different. It is a crucial tool that enables a complex relationship to be managed effectively – not measuring for measurement sake. The Relationship Managers who know their organizations and the staff should nominate those who will take part. A comprehensive sample of knowledgeable people drawn from a wide range of functions and organizational levels will bring in the best quality data. Remember, those 'at the coal face' will have valuable knowledge and experience. This is not sampling research but instead a 'self-selected census'. As long as the right people with good knowledge are chosen to take part, the number of people surveyed is immaterial.

Rules of Engagement All questionnaire respondents must be assured total anonymity and confidentiality in order to secure their open, honest, and complete participation. There is no data collection by workshop. We consider this as the most unreliable way of collecting objective data. If you are sitting next to your boss or your opposite number from the other organization, you are likely to hold back from expressing your true views because you don't want to open yourself up to career-limiting honesty, organizational embarrassment, or spoil relationships with those you interact with on a regular basis.

How to Survey? This step minimizes the effort required while generating true perceptions of clear, important collaborative relationship issues. The survey should consist of a relatively short questionnaire that is designed to show what is happening at critical points in the relationship that occur between the organizations. The questions should focus on the relationship management drivers. They therefore do not need to vary, apart from context, regardless of industry sector or setting. The questionnaire should not take more than 15 minutes to complete. It should ask for 'agree/disagree/don't know' answers to simple questions that require some thought and force the respondent to clarify his or her ideas. As is usual with surveys, there should be a text entry box at the end to allow respondents to freely contribute anything they think is relevant. In one live survey we carried out, a free text entry amounted to 8 × A4 pages when printed out! It was clear that we were the first people to ask the respondent his opinion.

Step 2 – Interviews

What People Say – Asking Why Focused interviews are used to add richness to the survey responses, and again, absolute confidentiality must be guaranteed. The purpose of the interviews is to provide the reasoning behind the survey results. Interviewees will be asked to comment on the joint, headline, traffic light results from the survey. For example:

- 'Why is this aspect so successful, and what is its impact on the joint enterprise?' 'What efforts have been made and could still be made to capitalize on this achievement?'
- 'Why is there such a difference of opinion between the organizations on this particular matter, and how does it affect the way the relationship operates?'
- 'They/you have indicated that this aspect is not performing well. Why?' 'What has been/is being done to overcome the issue?'

As long as the right people are selected, that is, those with the roles and seniority who have an overview of both strategic and tactical parts of the relationship, only a small number of interviews will be needed. Moreover, each session need not take more than an hour to complete. It is preferable to carry out interviews remotely/over the phone. This is less intrusive and time-consuming for the interviewee as well as allowing him or her to contribute without pressure or distraction. It also enables the interviewer to concentrate more fully on making notes of the key statements made by the interviewee without the need to record and transcribe the conversation. It is likely that quotations from both the questionnaire comments and the interviews will be used to add impact to the assessment findings. These must be sanitized to ensure that it is not possible to identify individual contributors or the firms they came from.

> It is important that the results of your study are promulgated to the senior levels in the company so that they will allocate more resources to improve the service.
>
> **Technical Services Manager, Raytheon**

Step 3 – 'Workshopping' the Way Forward

Call to Action The results of the surveys and interviews should be considered in a short sharp workshop facilitated by the Relationship Managers. It is designed to achieve four things:

- Endorse the survey and interview findings
- Add any further richness to the information revealed
- Decide and prioritize any actions that need to be taken
- Agree and resource an implementation plan

The overall focus must be joint value-seeking:

- How do we capitalize and further exploit those things we are doing well?
- How do we eliminate waste and turn-round dysfunctional processes?
- How do we put in place joint management that aims to continuously improve relationship performance?
- How do we ensure we seize and capitalize every new business opportunity that presents itself to the alliance?

The attendees from the participating organizations will include the senior sponsors, often at director level and above; the Relationship Managers who will take the action program forward; and departmental managers and key staff who will be affected. The group must not indulge in finger-pointing and laying the blame at anyone's door because this is counter-productive. The team must instead face its challenges as a team, and low-performing areas must be turned round or re-deployed to those who can contribute greater value. It is usually productive to re-state and confirm the original alliance value objectives in order to re-establish the motivation for collaboration. Often these will have been overlooked or forgotten in the 'white heat' of trying to establish and make the relationship work.

> I would like to repeat this process over time to see how perceptions change as a result of our joint efforts.
>
> **Senior Project Manager, Rolls Royce Aero-Engines**

Implementing Change

The implementation plan must be integrated with the joint management of the relationship so that not only does it ensure those things that need doing get done, but also that aspiration targets can be set and achieved. This is what creates continuous improvement because reactive change initiatives only result in a temporary hold up in relationship performance decline. In Figure 8.3, we show how relationship assessment dovetails with change implementation into a continuous improvement cycle. All changes must be institutionalized by its incorporation in policy, operational processes and practices, staff training, and internal audit. We usually see this happening 'in-house', but it is also critical to apply it to the joint management of collaborative relationships.

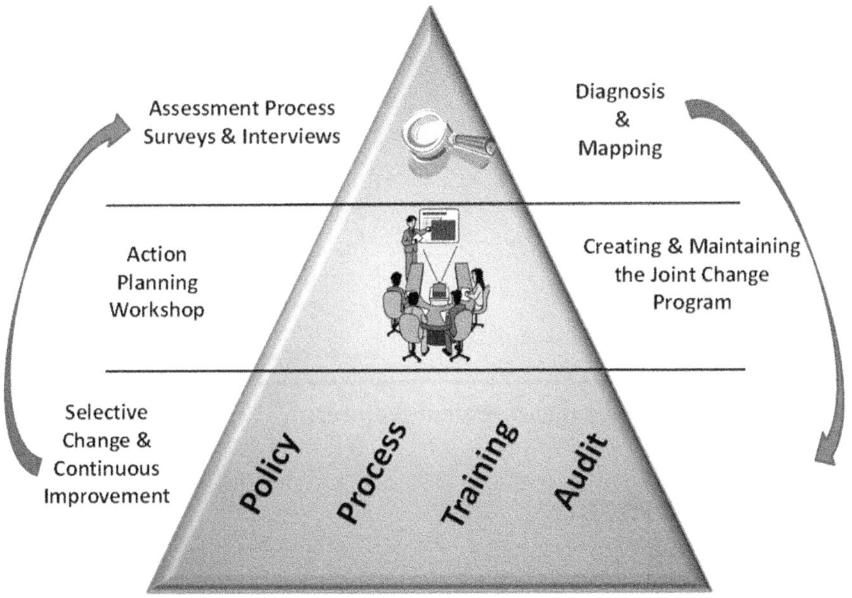

Figure 8.3 The collaborative change cycle.

The big shift in focus will be away from price and compliance and toward how organizations address innovation and change through their agreements.

Bill Huber, Information Services Group (IACCM – The Future of Contracting)

Relationship Performance Measurement Summarized

In Figure 8.4, we summarize the key features of the performance measurement process. The overall aim is to capture and utilize reliable data in a rigorously repeatable method. Moreover, because the data are based on a standard model, it means that the collaborative performance of organizations can be compared and benchmarks used to make strategic and performance judgments. Thus, senior managers gain a powerful governance tool in a key aspect of business that is not available using traditional methods.

Initial Joint Briefing:
- Attended by senior sponsors and managers
- Set/confirm joint relationship objectives
- Identify Relationship Managers who will coordinate the appraisal within their organizations
- Agree assessment process and project plan

Survey:
- **Who:** a 'self-selected census of key informants', as many knowledgeable staff as possible across relevant roles, functions and management levels

Interviews
- Small number of senior individuals who have a broad view of the organization and the relationship
- The interviewees are asked to verify and comment on the high level survey results
- Key statements recorded, sorted, analyzed, categorized

Action Planning Workshop:
- Attended by Relationship Managers and senior managers
- Confirm and comment on the survey (what's going on) and the interview (why it's going on) findings
- Generate logical objectives and actions
- Agree a prioritized, resourced action plan

Figure 8.4 Key performance measurement features.

Chapter 8 Worksheet

The purpose of this worksheet is to understand what is going on in your relationships that affect its performance.

Activities

- Identify the regular and repeatable methods that are used to uncover the root causes of what is going well and what is not going well within and across the relationship, for example:
  - Management reviews
  - Regular team and personnel interactions
  - Formal problem handling
  - Confidential issue reporting
  - Benefits tracking
- Compare what you have found with the principles described in this chapter
- Draft a new system for collecting and analyzing relationship performance data
- Run a pilot project to test the system with the co-operation of your partner

Questions

Are these measures effective? Do they reveal a clear, reliable picture of what is happening?

Do you need a different set of metrics? How will you go about introducing them?

> Inappropriate performance measures that determine rewards are preventing us from delivering to expectations.
>
> **Relationship Manager, Wipro**

Chapter 9

Performance Management – If You've Measured It, You Can Now Manage It

Applying Performance Measurement

Now that you have understood what relationship performance measurement is all about, how do you apply it in your situation in order to maximize collaborative value creation? In this chapter, we run through the four scenarios of a generic, structured approach to using these measurements in order to manage performance and maintain forward momentum. At each stage, we include an illustrative case study.

- Scenario 1 – Achieve high-level understanding of each existing relationship
- Scenario 2 – Create and analyze the relationships portfolio 'map' in order to understand where good and poor practice lies
- Scenario 3 – Carry out a deeper diagnosis of problem relationships
- Scenario 4 – Review on an annual basis to ensure continuous improvement

We also take a brief look at two other situations:

- To assess your organization's readiness to take advantage of a new opportunity that would benefit from collaborative working

DOI: 10.4324/9781003221340-9

- To assess the performance of a group of partners in an alliance or consortium

Moreover, we open the door to thinking about your own preparedness to collaborate and to work within a group. Whether you are in a one-to-one or one-to-many relationship, performance management plays an important part in your enterprise relationship governance arrangements. It is most likely you will find yourself in one of these situations where an objective view of performance is crucial:

- Kick-starting a management review and performance improvement program
- Initiating a relationship intensification exercise, e.g. at a new phase point
- Carrying out a strategic appraisal of key relationships
- Re-negotiation of contracts, especially evaluating in-house bids, assessing facilities management, or outsourcing arrangements
- Managing an ongoing, in-depth review and improvement program

> Performance requirements are very clearly stated in our contracts. Not sure that the measures actually roll down to individuals, hence there is uncertainty over how each person contributes.
>
> **Commercial Manager, Metronet Rail**

Scenario 1: Quick Look

This provides a strategic performance overview of each alliance relationship to identify the salient strengths and weaknesses of operational interactions. It will include:

- Initial briefing of the Relationship Managers

Figure 9.1 is an example one-page set of traffic lights that resulted from surveying the views of a pair of partner organizations.

The percentage satisfaction scores with the relationship are shown in colored traffic lights and numbers. It's possible to compare the respective perceptions quite easily by focusing on the colors. The main measures are listed on the left as are the softer aspect such as Trust and Commitment.

	Joint	Customer	Supplier
Main Measures			
Overall	73	79	66
Innovation	81	92	71
Investment	71	86	56
Communication	62	74	50
Operations	67	63	70
Value	82	79	86
Additional Characteristics			
Long Term Orientation		72	72
Interdependence		100	50
C3 Behavior		77	48
Trust		100	33
Commitment		92	75
Adaption		69	74
Personal Relationships		100	44

- **Innovation** – the leap of faith, being creative, flexible and resilient
- **Investment** – alignment of objectives, investment in people, know-how, infrastructure and management effort, and long-term vision
- **Communication** – open and transparent, frequent and extensive, learning, planning and anticipating
- **Operations** – focusing on service and product delivery, lowering joint costs and risks, building trust
- **Value** – perceived and actual benefits, satisfaction

- **Long-term Orientation** – encouraging stability, continuity, predictability and long-term, joint gains
- **Interdependence** – loss in autonomy is compensated through the expected gains
- **C3 Behavior** – Collaboration, Cooperation, Co-ordination, joint resourcing to achieve effective operations
- **Trust** – richer interaction between parties to create goodwill and the incentive to go the extra mile
- **Commitment** – the relationship is so important that it warrants maximum effort to maintain it
- **Adaption** – willingness to adapt products, procedures, inventory, management, attitudes values, and goals to the needs of the relationship
- **Personal Relationships** – generating trust and openness through personal interaction

Bandings		
0-49:	Red	Urgent Action Required
50-59:	Amber	Corrective Action Required
60-74:	Amber / Green	Corrective Action Recommended
75-100:	Green	OK Unless High Priority

Figure 9.1 Performance at a glance.

CASE STUDY: ADAPT, LEARN, AND PROSPER

The partners in Figure 9.1 were a pair of SMEs, a specialist, Electronics Company, and its parts supplier. It can be seen that neither was satisfied with their *performance*, although the customer was more dissatisfied than the supplier. Their 20-year relationship had matured into a comfortable 'marriage' where conversations had become mundane grumbles and the innovative spark had died. They had both lost track of each other's changing and improving capabilities and due to failure to capitalize on the possibilities, they were losing their competitive edge. The assessment came as a wake-up call and as can be seen in the bottom line figures below, the partners were able to affect a transformation. Good relationship management from the start would have prevented the performance degradation from occurring.

- Regular Planning Meetings defined joint production schedules
- Joint product reliability action saved £50,000 per year on in-house testing and £90,000 in parts holdings
- Supplier involvement in new product design brought lower cost, improved design reliability, better asset availability, more functionality, and shortened time to market from 5 years to 1 year
- Customer updated MIS with new, integrated system to improve asset control, marketing, and requirements forecasting
- Over 3 years, Customer Revenue was up 38.5% per year and Supplier Revenue was up 35% per year

We are now concentrating on solving the issues rather than shouting at each other.

CEO, Electronics Parts Supply Company

Scenario 2: Portfolio Perspective

Where you have a number of key relationships, such as with your main customers or suppliers, it is essential to understand their strategic importance and to have an overview of the performance of the portfolio. This can be created using a summary report of the main management points from a program of 'quick look' assessments. This facilitates strategic management of the portfolio including identification of common factors to enable putting in-train development programs and setting targets for improvement.

Figure 9.2 is a view of 57 relationships between a customer and his main suppliers. The top-level performance figure from each relationship is plotted showing the positive and negative scores (a relationship having an overall satisfaction score of 75% is shown with a green vertical bar for 75% and a red vertical bar for 25%). Highlighting the green scores identifies where the good practice can be found, and the red shows where help is needed the

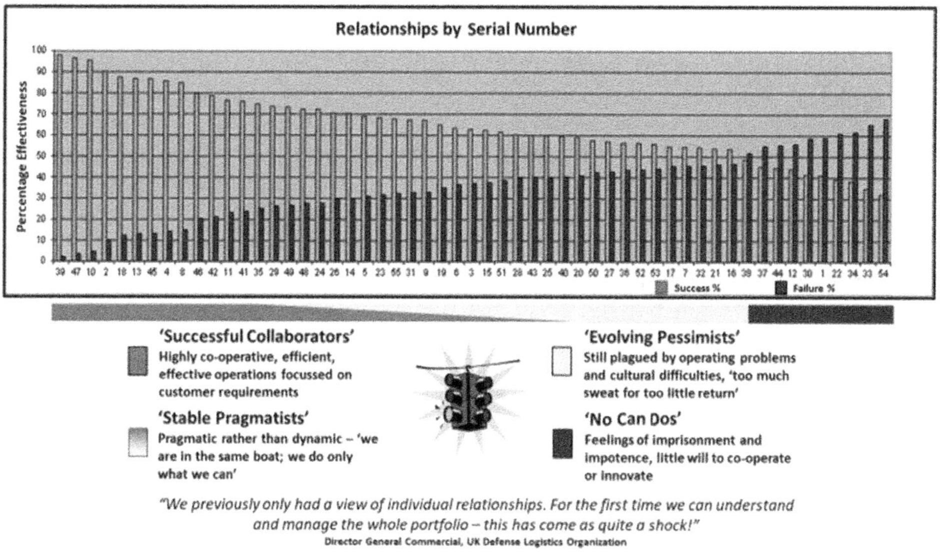

Figure 9.2 Portfolio at a glance.

most. The customer's portfolio comprised over 200 strategically important relationships. This sample included small, medium, and very large alliances that were felt to be representative of the complete set.

CASE STUDY: A NEW GOVERNANCE OPPORTUNITY

In this case, the customer's head office working with their own and the suppliers' managers generated a catalogue of best practices and practitioners from its best relationships. They then ran a couple of projects using these resources to support and mentor a very poor and a medium relationship. The aim was not only to change their 'game' but also to pilot an approach that could be rolled out across the portfolio with the intention of standardizing the approach to relationship management. This plan also allowed the targeting of corporate-level resources at prevalent issues such as complex technological environments (engineering), retaining and motivating staff with critical skills (HR), standardizing collaborative contracts (commercial), improving knowledge-sharing (IT), and consignment/item tracking (supply chain).

Finally, the Group CEO met with the industry steering group's leaders and decided that the portfolio perspective provided them with a new governance opportunity. It would allow them to set performance targets for the portfolio as a whole that would cascade down via corporate directors to the individual project relationships.

> This is great; for the first time I can now talk with my opposite numbers with real facts about strategic business performance at my fingertips.
>
> **Group CEO, Defense Equipment Support Organization**

Scenario 3: Strategic Partnership

This is an in-depth diagnostic that uses a combination of relationship measurement and relationship analysis. It provides metrics that reveal the current position including clear indications of strengths and weaknesses. Analysis builds upon this information to generate an in-depth understanding of drivers behind the metrics. This enables the partners to decide on new objectives and choose the action necessary to reach these goals. This strengthens the relationship by mobilizing joint support for change and making success more achievable.

The process includes:

- An initial briefing of the Relationship Managers
- An online survey of all knowledgeable staff
- An interim report
- Telephone interviews of selected staff
- A facilitated workshop for alliance Relationship Managers and their teams to agree the key performance factors in the relationship
- An action plan for improvement

CASE STUDY: CROSSED WIRES

This is a long-standing relationship of 20 years that has in the past provided the partners with great considerable value and satisfaction. However, in recent times, changes in the way the relationship has been managed by Firm A has resulted in inefficiency and frustration. It is apparent in Figure 9.3 that Firm A feels very strongly that it is trapped in a 'loveless marriage' and the benefits have lost their luster. Overall, however, there is a feeling that the relationship has a future and 'divorce' is not the answer. It turned out that despite satisfactory operational performance, communications between the partners were poor with format, frequency, and method problems resulting in 'crossed wires' and one particularly frustrated partner. It only took a joint review and then re-alignment of the

Main Measures	Joint	Firm A	Firm B
Overall	68	62	74
Innovation	60	52	68
Investment	63	61	65
Communication	77	75	79
Operations	60	52	68
Value	79	69	90

- **Innovation** – the leap of faith, being creative, flexible and resilient
- **Investment** – alignment of objectives, investment in people, know-how, infrastructure and management effort, and long-term vision
- **Communication** – open and transparent, frequent and extensive, learning, planning and anticipating
- **Operations** – focusing on service and product delivery, lowering joint costs and risks, building trust
- **Value** – perceived and actual benefits, satisfaction

Additional Characteristics	Firm A	Firm B
Long Term Orientation	81	80
Interdependence	72	86
C3 Behavior	67	67
Trust	63	82
Commitment	66	81
Adaption	70	90
Personal Relationships	33	56

- **Long-term Orientation** – encouraging stability, continuity, predictability and long-term, joint gains
- **Interdependence** – loss in autonomy is compensated through the expected gains
- **C3 Behavior** – Collaboration, Cooperation, Co-ordination, joint resourcing to achieve effective operations
- **Trust** – richer interaction between parties to create goodwill and the incentive to go the extra mile
- **Commitment** – the relationship is so important that it warrants maximum effort to maintain it
- **Adaption** – willingness to adapt products, procedures, inventory, management, attitudes, values, and goals to the needs of the relationship
- **Personal Relationships** – generating trust and openness through personal interaction

Bandings		
0-49:	Red	Urgent Action Required
50-59:	Amber	Corrective Action Required
60-74:	Amber / Green	Corrective Action Recommended
75-100:	Green	OK Unless High Priority

Figure 9.3 Crossed wires.

information channels in this alliance to overcome the issues. More importantly, the diagnostics exposed an unexploited opportunity to leverage greater value from the data flows about customer demand trends.

> We get extremely frustrated with this lack of passion and the inability of our contacts to 'fix' issues, i.e. supply chain deficiencies over the years.
>
> **MD, Specialist Confectionery Manufacturer**

Scenario 4: Regular Reviews

We strongly suggest that in order to achieve continuous improvement, each key relationship that has been measured using Scenario 1 and Scenario 2 be re-measured regularly or at least annually so that performance can be monitored and new objectives set. Normally only a Scenario 1 quick look is needed unless a strategic refocusing of the relationship has taken place.

Are You Ready to Collaborate?

When an opportunity arises that would bring greater benefits by partnering, then it is important to understand your organization's readiness and capability to partner effectively. For instance, are your processes joined up in order to effectively service your relationships, and this includes a management focus that encompasses the joint enterprise perspective? Scenario 1 quick look can be carried out as a self-assessed exercise. If this highlights issues, then Scenario 3 strategic partnership assessment can be applied to provide grounding for an internal change program.

Alliances and Consortia – Group Working

So far we have described performance measurement for individual companies working with a single partner. However, more and more companies are joining together in groups, networks, consortia, alliances, and complex supply chains. The TOM applies equally to these joint enterprises which require enterprise-wide performance management. To this end, Scenario 1 quick look and Scenario 3 strategic partnership should be applied in multi-party format.

CASE STUDY: CHILDREN IN CARE, JOINING THE DOTS

A multi-agency local authority task force was set up to operate a strategy for children in care. Difficulties were being experienced in achieving joined-up working. The main operation was only working successfully 70% of the time, which was not good when it impacted the lives of vulnerable young people. There was complacency, distrust, and despair in the collaborative. It was clear that some members believe that the implementation would either fail or not achieve its objectives. See Figure 9.4.

	Alliance	Firm A	Firm B	Firm C	Firm D	Firm E	Firm F
Main Measures							
Overall	62	62	74	76	79	30	48
Innovation	59	52	68	72	79	41	46
Investment	58	61	65	77	75	17	53
Communication	71	75	79	78	84	54	57
Operations	53	52	68	70	73	27	27
Value	67	69	90	83	89	14	57
Additional Characteristics							
Long Term Orientation		81	80	70	79	0	75
Interdependence		72	86	79	68	25	25
C3 Behavior		67	67	81	86	50	51
Trust		63	82	96	57	0	25
Commitment		66	81	75	73	13	58
Adaption		70	90	79	90	17	42
Personal Relationships		33	56	83	55	58	61

Bandings		
0-49	Red	Urgent Action Required
50-59	Amber	Corrective Action Required
60-74	Amber / Green	Corrective Action Recommended
75-100	Green	OK Unless High Priority

- **Innovation** – the leap of faith, being creative, flexible and resilient
- **Investment** – alignment of objectives, investment in people, know-how, infrastructure and management effort, and long-term vision
- **Communication** – open and transparent, frequent and extensive, learning, planning and anticipating
- **Operations** – focusing on service and product delivery, lowering joint costs and risks, building trust
- **Value** – perceived and actual benefits, satisfaction

- **Long-term Orientation** – encouraging stability, continuity, predictability and long-term, joint gains
- **Interdependence** – loss in autonomy is compensated through the expected gains
- **C3 Behavior** – Collaboration, Cooperation, Co-ordination, joint resourcing to achieve effective operations
- **Trust** – richer interaction between parties to create goodwill and the incentive to go the extra mile
- **Commitment** – the relationship is so important that it warrants maximum effort to maintain it
- **Adaption** – willingness to adapt products, procedures, inventory, management, attitudes values, and goals to the needs of the relationship
- **Personal Relationships** – generating trust and openness through personal interaction

Figure 9.4 Alliance performance at a glance.

Operational and process problems had a negative impact on the way people were interacting (see Figure 9.5). There were, however, a number of green scores that indicated the desire to make the collaborative perform better. It was recommended that they:

- Mapped the end-to-end processes for each service
- Reviewed the performance requirements of each process
- Reviewed the staff training and policy needs for each process
- Established a process manager for each who would be the main point of contact for its operational delivery
- Upgraded the information flows so that all members received timely, accurate communications
- Established a regular alliance management group with its own terms of reference and standard agenda
- Introduced a team-building program
- Carried out further Alliance Performance assessments annually

Operations

Focusing on service and product delivery, lowering joint costs and risks, building trust

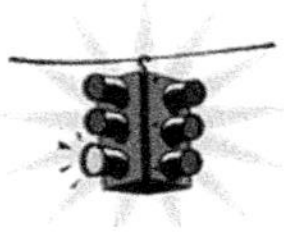

		Alliance	Firm A	Firm B	Firm C	Firm D	Firm E	Firm F
Q4a	The quality of the contract outputs e.g. consistent product/service quality, fulfilled on-time orders is entirely satisfactory	52	83	75	33	86	0	33
Q4b	The quality of service e.g. billing, prompt payment, administration, delivery is entirely satisfactory	52	75	67	45	100	25	0
Q4c	The alliance is characterised by a continually improving product quality philosophy	72	63	100	67	86	67	50
Q4d	Problems are solved in a joint, open, constructive manner	47	25	78	92	87	0	0
Q4e	Such is the goodwill in the alliance, the partners would willingly put themselves out to adapt to each other's changing requirements	40	13	22	83	21	50	50
Q4f	The alliance partners trust each other to act in their best interests	43	38	78	92	50	0	0
Q4g	The responsibility for making sure the alliance relationships work is shared by the partners	74	50	100	92	93	75	33
Q4h	The alliance partners provide each other with useful cost reduction and quality improvement ideas	59	0	14	67	55	50	67
Q4i	The alliance partners are totally open and honest with each other	42	38	75	75	64	0	0
Q4j	The alliance partners always do what they say they will do	46	38	67	50	86	0	33

Figure 9.5 Disjointed operations.

We send out minutes and papers to them and expect responses by the due date. Sometimes it takes weeks to get an answer; their key people are never available.

Social Services Manager, City Council

Relationship Performance Management Benefits

In Chapters 8 and 9, we have described a simple, practical way of using performance measurement specifically designed for managing collaborative relationships. We presented the principles of collaborative performance management and how they fit into the TOM. We then offered a step-by-step guide to our assessment process, followed by some pointers on how it can be used depending on your circumstances. Here are the key benefits of effective relationship performance management that we have observed in our extensive experience of working with organizations of all shapes and sizes in both the public and private sectors.

- Transforms perceptions by promoting understanding of the way that partners view each other, thus promoting trust and co-operation
- Allows joint change programs to be anchored in the aims and objectives and measured performance of the relationship
- Allows strategic high ground to be taken to strengthen relationships so that:
  - Fire-fighting becomes unnecessary
  - Continuous improvements in effectiveness, including meeting international standards, are achieved
  - Partners can concentrate on the long-term value for money rather than satisfying contract 'small print'
- Becomes the catalyst for defining joint opportunities to increase revenue and shareholder value
- Increases customer satisfaction from better product/service quality and delivery
- Strengthens the bond between partners and enhances their ability to innovate
- Reduces administration and production costs and risks
- Bridges the hidden gaps in teamwork and increases productivity, through more cohesion, integration, fulfillment, and increased transparency

- Builds joint capability to seize future business opportunities
- Reduces costs without diminishing efficiency or jeopardizing relationships
- Eliminates the risk that a weak relationship will let you down at a crucial time
- Exposes in detail the strengths and weaknesses, areas of inefficiency, and where wastage may be occurring in the functions and processes within and between the alliance members
- Enables effective relationship management based on objective information
- Makes sense of 'messy'/complex situations which allows managers to focus on the essential success factors
- Highlights counter-productive activities and describes actions to turn them round
- Highlights positive activities and describes actions to enhance and spread good practice
- Mobilizes joint support and creates a common foundation for change
- Promotes innovation and continuous improvement
- Supports proactive governance and is a key enabler of Enterprise Relationship Management

> A strong indicator that you have a good partner is the confidence that they will do what they say they will do, they can be trusted to share in an open and transparent way, and we will both be comfortable in a situation where we are interdependent.
>
> **Director, Siemens Power Engineering**

Chapter 9 Worksheet

The purpose of this worksheet is to understand your relationship performance management arrangements and how they can be improved.

Activities

- Having identified how you measure relationship performance, with your partner consider how it feeds into a joint, ongoing cycle of change and improvement

- List the positive benefits you have found
- List the gaps, their impact, and reasons
- Describe how your TOM is continually improved as a result

Questions

Are you confident in your relationship performance management arrangements?

Do you compare your relationship performance management across your portfolio of partnerships in order to implement systemic improvements?

> We are starting to face up to performance issues at our regular meetings. In the past, mutual defensiveness has got in the way of making improvements. This is a very hard matter to tackle.
>
> **Senior Manager, Gas Turbines Manufacturer**

Chapter 10

Managing Exit – All's Well That Ends Well

Every successful business relationship will eventually come to an end; this is a natural progression. Just as in other aspects of relationship management, the exit phase needs to be carefully planned and its execution managed. This will ensure that a satisfactory outcome for all stakeholders will be achieved including the retention of goodwill and therefore the opportunity to do business in the future. A key benefit is learning from experience and improving relationship management capability.

What Is Exit?

Exit is the process of ending a collaborative relationship. It must ensure that if an alternative provision is required for any reason, an adequate level of support can be maintained without disruption to operations. As the relationship has been collaborative, the exit is likely to be complex and require considerable planning and effort. Significant intellectual property rights will often be involved as well as those investments that have been made and used by the partners such as skills, assets, and infrastructure. The impact on the up-stream and down-stream members of the supply chain should not be forgotten, including of course the all-important end customers. The focus of the partners is around working collaboratively now and in the future so that on exit both parties are happy that they can work together again to maximize future opportunities.

DOI: 10.4324/9781003221340-10

Exit Scenarios

There are likely to be three different situations where the relationship is no longer needed:

- **Mutual Agreement**: Where, for example, the requirement for the product or service declines or ceases to exist because of market conditions, or agreed time limit as in the case of a completed project or change in regulatory or other external conditions is warranted
- **Unilateral Decision**: Where, for example, a partner decides to pursue a better offering from a new partner
- **Non-performance**: This arises as a result of action on one side or the other that triggers a breakup of the relationship. For example, culture mismatch could become serious enough to make an exit necessary, as could a failure of effective enterprise relationship management or under-performance. Alternatively, it could arise where one or other of the partners ceases trading, is taken over, or cannot fulfill its contractual obligations

Within these there are two types of exit. The principles – to achieve a smooth and efficient termination – are the same, but the operational implications are different:

- **Shutting down**: Closing down an operation gracefully usually when the partnership is dissolved and the joint enterprise ceases to exist
- **Transitioning**: Transferring part of/the full operation to an alternative partner. In this instance, the enterprise needs to continue with as little disruption to end customers as possible during the transition. There may be times when a new partner has to be found and interim arrangements will need to be put in place to sustain the operation

The Exit Landscape

At the outset in a joint enterprise there will be enthusiasm and euphoria. But the objective of the venture to generate clear benefits for all concerned throughout its life should always remain center stage. Therefore, at the earliest point, even before the deal is sealed, it is essential to think jointly, openly, and honestly about potential exit scenarios. During the life of the enterprise, partners may come and go and finally the relationship will be dissolved. It is vital that value continues to be captured and exploited to the full, and thus damaging litigation is avoided. Although it may seem perverse, maintaining a constant management focus on exit planning is an essential component of joint enterprise success. Accordingly, an exit plan should be jointly developed to prepare for further relationships beyond the duration of the current contract.

Why Plan for Exit: What Could Possibly Go Wrong?

It may surprise you to know that very few joint enterprises plan for the end of the relationship, and as a result many end badly in a courtroom with expensive squabbles over who does and owns what and who gets the spoils.

> Forty percent of strategic alliances fail within 5 years. They fail to pick the right partner, they don't agree goals up front, there is a lack of relationship and chemistry, communications are inadequate and erratic, and there is too much interference by parent companies.
>
> **Elix-IRR partners LLP, 2011**

This may be good for the lawyers, but it is often disastrous for the end customers. Here are some salutary case studies that illustrate some of these situations.

CASE STUDY: ASTRAZENECA UK LIMITED VS. INTERNATIONAL BUSINESS MACHINES CORPORATION, 2011

This partnership ended up in court because neither party had planned effectively for winding up their relationship. The following is the judgment handed down in the case.

> This case demonstrates the importance of ensuring that exit management provisions receive careful attention from the parties and their legal advisers during contract negotiation. Failing to consider the parties' obligations in the event of termination can lead to

considerable uncertainty and increased costs. Parties embarking on high-value, complex IT outsourcing projects should focus their minds on what they want to happen should their commercial relationship come to an end (this should not be an afterthought) and should resist the temptation to be swept up in the deal which is being done.

The opportune time to negotiate such matters is pre-contract when the parties are on a stable footing and their relationship is likely to be at its strongest. It is important that the parties agree a set of guiding principles for managing exit, and a solid exit management framework should be included in the agreement. Fine-tuning of the exit management process during the term of the contract, perhaps through formal annual reviews of any exit plan, may also provide the parties with a useful opportunity to add detail to an agreed framework and address any unforeseen issues which may have arisen. However, attempting to negotiate exit processes following termination, perhaps in acrimonious circumstances, is unlikely to be productive or to result in a favorable outcome for either party.

Judgment, High Court of England and Wales 306

CASE STUDY: AIRWAVE SOLUTIONS AND THE UK'S EMERGENCY SERVICES

Airwave Solutions was commissioned by the UK Home Office in 2000 to create the TETRA emergency services radio network using mobile phone technology. Unfortunately, this relatively small supplier could not afford the investment needed for next-generation technology when its network technology had become obsolete. In 2010, the new Conservative/Liberal Democrat coalition government decided that a major upgrade was needed to take advantage of new developments such as 4G. The Home Office decided to replace TETRA Net with a brand new alternative at a cost of approximately 4.8 billion GBP. Airwave Solutions initiated legal action in 2015 against the Home Office over its decision to award part of its Emergency Services Network contract to EE but then went out of business and had to be rescued by Motorola to secure the transition to the new system.

Tetra-Applications.com (weblog)

CASE STUDY: BT AND AT&T'S CONCERT JOINT VENTURE

Headline: 'Concert, a $10 billion joint venture of AT&T Corp. and British Telecom that focused on offering a broad range of global communications migrating customer networks to an IP-based environment, is calling it quits'.

After only 2 years, in 2001, the arrangement was wound up after losing about $800 million annually. The two companies wrote off $7 billion in charges with AT&T taking the brunt of the damage, writing off $5.3 billion. BT lost $1.7 billion and under the breakup agreement 2,300 of Concert's 6,300 employees lost their jobs with BT paying $350 million in redundancy costs. The CEOs stated they deliberately did not address exit strategies 'because this cannot fail'. It did fail and it took years to untangle the resultant mess.

Tim Cummins, CEO, IACCM

Strategy – Unwinding the Joint Enterprise

Because the relationship has been collaborative, the exit is likely to be complex and require considerable planning and effort. Significant intellectual property rights will often be involved as well as those investments that have been made and used by the partners such as skills, materials, and infrastructure. The impact on the up-stream and down-stream members of the supply chain should not be forgotten. If at all possible, the exit should be managed so that the impact on both organizations and their stakeholders is minimized, and future collaboration between the partners is not jeopardized.

Managing for Exit

In Chapter 1, we have described a TOM for managing collaborative enterprises. Exit permeates every aspect of it. There are three main components that must be in place:

- **The Relationship Manager**: They play a pivotal part in ensuring the smooth running and therefore the performance of business-to-business relationships. Thus, the RMs will play a key role in managing the exit process throughout the life of the joint enterprise. See Chapter 3

- **ERMP**: The source of all relationship information necessary for managing exit will have been recorded in the appropriate sections of the ERMP. See Chapter 4. For example, in the commercial section you are likely to find the reasons for termination. Pertinent exit information will have been kept updated during the life of the relationship
- **The Exit Plan**: This will contain up-to-date information which will be used for developing a detailed project plan to initiate and implement an orderly exit. It will include such things as exit scenarios, dispersal of intellectual property rights, and other joint/shared assets. It will also consider impacts on all stakeholders including end customers and suppliers and how to mitigate that impact

The Exit Management Process

We now describe in detail how exit planning fits into all the life of a relationship.

Laying the Foundations for Exit During the Decision Phase, we go through a number of steps in order to choose the right partner. See Chapter 5. There are three main areas of focus: attitudes to risk, building the business case, and reviewing the way that the relationship will be managed. Integral to this is the need to consider and draft the exit plan. If you cannot clearly see a satisfactory conclusion to the joint enterprise, then the decision to continue would be in doubt.

Attitudes to Risk One of the first steps is to discuss the potential partners' attitudes to risk. This will identify those conditions under which the joint enterprise will terminate and the likely responses including resources needed to secure an orderly conclusion to the relationship. It is essential to ensure that a potential partner has a realistic grip on the risks they face in their business, their exit planning, and their understanding and willingness to bring this capability to the joint enterprise. It is crucial that both partners are as open and honest as possible. It is inevitable that more attractive options will arise in the future which will tempt a partner to leave the relationship. These scenarios need to be identified as part of the planning process in so far as they can.

Building the Business Case Having aligned and prioritized their objectives, the prospective partners need to explore the business case in order to discover the practicality and extent of the value of the proposed collaborative opportunity. This process should consider the implications for an exit. In particular, special attention should be devoted to the fact that the partner operations will be intertwined and exit will always be more difficult and costly. The business case should also include the situation when transfer of operations from one partner to another is necessary.

Case Study: Locked-In Partners A major oil and gas company contracted with an FM supplier for all its site support, including catering facilities. The FM supplier constructed and owned those facilities. When the oil and gas company wanted to replace them as a supplier, they essentially could not because a new supplier would be forced to build new facilities. So the decision to have the FM supplier build those facilities was attractive at the beginning, but essentially prevented exit. At the very least, it required agreement on a purchase price for the facilities if exit ever occurred.

Tim Cummins, CEO, IACCM

Exit and Relationship Management Prospective partners will consider at a high level each component of the ERMP in turn and insert any available information and ideas about how the joint enterprise will be managed. Consideration of the exit arrangements is designed to highlight the ramifications of dissolving the partnership so that this can be included in the overall decision on whether the enterprise is viable.

Checklist for Exit Planning in the Decision Phase

Below is a summary of points that need to be considered when making a decision to collaborate:

Identify potential partners' attitudes to joint risks that could result in exit:

- Conditions under which the joint enterprise will terminate
- Likely responses including resources and costs needed to secure an orderly conclusion to the relationship

Ensure that a potential partner has a realistic grip on the risks:

- In their business
- In their exit planning
- How they would bring this capability to the joint enterprise

Draft the high-level exit plan for inclusion in the outline joint business case considering:

- The most likely exit scenarios
- The impact on partners' organizations and their stakeholders
- Intellectual property and possible joint/shared assets
- Costs and resourcing
- The impact on customers and the up-stream and down-stream members of the supply chain

Review the potential management implications of exit on:

- Operations
- Infrastructure
- Personnel
- Finance
- Commercial
- IPR
- Supply chain
- IT
- End customers
- Stakeholders
- Exit team and resources
- Risk

Maintain the exit plan throughout the life of the joint enterprise.

Transitioning to Operations

The potential exit implications of the proposed enterprise have been outlined and discussed as part of selecting a partner, drafting a business case, and exploring collaborative management arrangements. These have formed an important part of the viability decision. These foundations are then used to create the policy basis on which to establish the new business. See Chapter 6.

Full Business Case This is the foundation for the collaborative project and includes the following exit plan elements:

- Exit scenarios
- Impact on
  - partners' organizations
  - stakeholders
  - customers
  - up-stream and down-stream members of the supply chain
- Intellectual property rights
- Assets

Commercial Agreement The collaborative commercial agreement will be future-looking and encourage creativity. It will be a flexible, proactive framework which will include exit provisions that support the principles specified within the full business case.

Business Plan This is the part of the business case that details resourcing, timescales, organization structure, responsibilities, and governance. The plan needs to take into consideration the potential implications of having to 'unwind' the operation at exit.

Exit Plan This is a strategic view that will contain all the information necessary to initiate the exit implementation project. Exit considerations run through the policy framework like a thread, as shown in Table 10.1. The details behind the exit plan will be contained in the ERMP. See Chapter 4.

Table 10.1 Exit and the ERMP

Exit scenarios	ERMP Risk Assessment section (Chapter 4)
Impact on partners' organizations and their stakeholders	ERMP Risk Assessment section (Chapter 4)
Intellectual property	ERMP Knowledge Management section (Chapter 4)
Assets	ERMP Asset Register section (Chapter 4)
Impact on customers and the up-stream and down-stream members of the supply chain	ERMP Risk Assessment section (Chapter 4)
Implementation resources	ERMP Organizational Arrangements section (Chapter 4) ERMP Communications section (Chapter 4) ERMP Personnel and Team Development section (Chapter 4)
Improving relationship management capability – learning from experience	ERMP Exit Arrangements section (Chapter 4)

Operations – Keeping Up-to-Date

Every joint enterprise, even if it is a defined project, will be subject to change and development. This is the nature of collaborative relationships. It is essential that the exit plan keeps pace with the tempo of the evolution. All changes must be documented in the ERMP. The relationship management process will be operated on two levels:

Strategic Management At the senior level, strategic reviews will examine performance of the past and present, ensure that the joint objectives are still appropriate, and look ahead to how the joint business will develop in the next period. They will also consider the effect of changes in the business, its markets and its environment, and their impacts on risk. See Chapter 7. As a result, the exit scenarios may need to be modified and the exit plan amended accordingly. They may also decide that the business, in part or in total, is no longer viable and that the exit plan should be invoked.

Operational Management At the tactical level, the RMs will meet regularly to review work schedules and progress to ensure that targets are being met. This involves meeting frequently to manage operations and less often to review the enterprise policy framework. See Chapter 7. They should be vigilant in spotting significant changes that are affecting operations and will impact the exit plan. They may be the first to realize that part or all of the enterprise should be shutdown.

Implement the Exit Plan

The joint enterprise 'board' having decided that a relationship should be terminated will task the RMs with carrying out the exit implementation project. We describe how the exit plan is turned into a project plan.

Managing exit implementation will involve the following steps:

- Issue the exit directive
- Define RM responsibilities
- Issue the Communication Plan
- Appoint a planning team
- Assemble the data
- Design the project plan
- Resource the project plan (people, money, equipment, and contracts)
- Run the project to successful completion

- Review and learn from experience

Empower the RMs The joint enterprise 'board' having decided that a relationship should be terminated will review the exit plan with the RMs. They will define the scenario, the objective, and the parameters to be used. They will pass authority for implementation to the RMs including the appropriate budget. Internal and external communication will be a key part of the project.

Do the Ground Work The RMs will gather a small expert design team to create the project plan. This team may need to include outsiders such as key members of the supply chain and other specialists. They will assemble the data starting out from the ERMP which includes the project policy and framework documents. Although these have been maintained throughout the life of the enterprise, it will be necessary to carry out a further review to ensure that the information is current in order to establish the project base line.

Create and Resource the Project Plan Next the expert design team builds the project plan. They will include the activities, priorities, timescales, and responsibilities. They will then resource the plan with people (managers, workers, commercial, legal, and technical), equipment/IT, outside support, and budget. A major component of the project will be an enterprise-wide communication plan which also plots information flows to external stakeholders and supply chain members. Governance arrangements will need to be made clear.

Work the Plan The expert design team will form the project management group under the leadership of the RMs. They will organize governance meetings where progress will be monitored, problems resolved, and serious issues will be reported to the joint board. A record of all management activities associated with working the exit implementation plan must be maintained, both as an audit trail and to enable the parties to learn from experience afterwards.

Review and Learn On completion of the project, a 'learning from experience' (LfE) report will be prepared by the RMs. Its purpose is primarily to improve the relationship management capability of the partners; however, business skills transfer is also important. It will ensure that the impact on all the organizations and their stakeholders is positive and the opportunity to do business in the future is not jeopardized. The review should consider what could have been done to avoid premature exit, the adequacy of the exit plan, and the management of the implementation project.

A copy of the report will be lodged in the ERMP, which as a shared resource will be available in a continued partnership. Where there are departing partners, they will take away the termination ERMP for their own capability development.

Checklist for Exit Implementation – Senior Management

- Decide to exit
- Review the exit plan with the RMs
- Issue a directive
  - Define the scenario, objective, and parameters to be used
  - Pass authority for implementation to the RMs
  - Provide a budget for the project
- Monitor progress
- Provide strategic guidance to RMs
- Manage high-level communications

Checklist for Exit Implementation – RMs

- Gather a small expert design team to create the project plan
- Assemble the data from the policy project framework documents in the ERMP
- Review and create the project baseline build project plan
  - Include activities, priorities, timescales, and responsibilities
  - Resource the plan with people, equipment/IT, outside support, and budget
- Create a communication plan to plot information flows
- Set up governance arrangements
- Form the project management group
- Meetings to monitor progress and resolve problems
- Refer serious issues to Senior Management
- Maintain project management records
- Prepare and distribute LfE report, include copy in ERMP

Exit Plan Guidance Every exit plan will be different, depending on the nature of the joint enterprise. Accordingly, it is impossible to provide a comprehensive guide; however, below are some aspects that are critical or are often overlooked or forgotten.

Operations Depending on your circumstances the plan will need to consider early access to at least some of the following:

- List of critical project lead times
- Assets/stock required to maintain supply, etc.
- Facilities that are unique to the task
- Expertise within personnel (either through direct transfer such as under a TUPE transfer or through skills transfer such as training/shadowing, etc.)
- IPR (licenses/design information)
- IT Systems, software, and licenses
- Data and key documents such as sales and marketing, financial, operational, logistics, personnel, orders, work in progress
- Data to enable other companies to realistically bid in a re-competition/partner selection exercise
- Processes and procedures
- Keep end customers informed
- Involve key supply chain members

Transitioning There are some additional factors that need to be considered when transferring part of/the full task to an alternative partner:

- Consider how much reliance there will be on the continued co-operation of the original partner and ensure how this will be affected when another partner is appointed
- An agreed plan is needed to cover the transition period
- Data required to resume/maintain the service by the next provider
- The existing partner and the new partner may need to share responsibility for execution of the service until transfer is complete
- Co-operation between all three parties will be essential to minimize any service disruption
- The commercial agreement must facilitate a smooth transition arrangement

Supply Chain Involve supply chain members (not forgetting key third party providers such as IT services) in exit planning.

Assets

- All assets must be recorded and classified according to ownership
- Assets including stock, IT, software, licenses, etc. regardless of their ownership must be available for as long as is necessary to execute exit
- There must be an agreed procedure – financial or otherwise, for distributing shared assets

IPR

- All IPR including licenses and design information and commercially sensitive information such as customer records must be recorded and classified according to ownership
- There must be an agreed procedure, financial or otherwise, for distributing shared IPR

Personnel

- Documentation to be prepared and made available by the partners includes data on personnel as defined by and constrained by employment legislation
- Ensuring that important skills and knowledge are identified and retained. This is particularly important where specialist personnel are key to the enterprise such as in technical areas

Finance

- Finalize accounts to obtain a clear picture of liabilities and profits which will facilitate equitable distribution of value
- Consider implications for taxes and pensions
- Consider accounting implications on individual partner organizations
- Ensure that sufficient reserves/insurances are held to carry out exit

Commercial

- The commercial agreement must fully reflect the partners' commitment to the exit plan
- Action on unilateral exit is necessary to mitigate the impact on the exited partner and reduce the likelihood of litigation
- Identify exit costs and resource requirements that could occur and agreeing parameters for provision
- Consider whether you need to form new supply chain relationships

End Customers

- Maintain a record of requirements including information, notice periods, and any action needed from them
- Keep in touch with them regularly to inform them of your exit arrangements and to reflect within it any changes in their requirements

- In some cases, ensuring continuity of service will be critical and financial compensation will not be acceptable

Stakeholders

- Remember to inform and consult with bankers, investors, shareholders, and other interested parties

Communication

- Manage internal (enterprise-wide) and external (the media and stakeholders) information flows

Exit resources

- Staff: roles, responsibilities, leadership, knowledge, experience
- Money: to hire in outside support, specialized equipment; transfer payments for assets, etc.
- Resources

Risk

- Good risk management including up-to-date scenarios is key to effective exit planning:
  - A scenario might identify what alternative partners exist, which represents the next best solution as an alternative, what actions would be required to effect reallocation of the work, and what impact this may have on continuity
  - Vulnerabilities within the lower tiers of the supply chain should be investigated
  - Awareness of individual partner risks and joint risks
- Consider most likely and most disruptive scenarios
- Risks can be mitigated by good relationship management and partnering behaviors:
  - Ensure that communication of task, changes, performance, and partner satisfaction is practiced
  - Define dispute resolution procedures in order to minimize relationship breakdown due to poor performance
  - Look ahead to possible or agreed organizational changes and review whether they could impact the relationship. There may be an opportunity to re-shape processes, deliverables, supply chain engagements in a manner to take advantage of value creation/continual improvement

Chapter 10 Worksheet

The purpose of this worksheet is to ensure that you have fully considered how, when, and why your relationship should end. Your aim is to guarantee that it will be amicable and efficient.

Activities

- At the beginning of this book you listed your main relationships; now review the exit plans for each
- Consider how workable they are
- Review them against the exit planning guidance above
- Identify the gaps and produce a plan to update them. They should be regularly reviewed jointly by the RMs

Questions

Have you or others in your industry had a bad experience over inadequate exit planning?

Are you satisfied with your exit planning?

> In 2008, the UK Department for Transport refuted claims that the Public Private Partnership was to blame for the collapse of Metronet, and that it was 'predominantly a corporate failure', with 'structural weaknesses [that] led to its own downfall'. In 2010, the House of Commons' Public Accounts Committee reprimanded the UK Department for Transport for its failure to heed UK National Audit Office warnings about the company's management and lack of termination planning. According to the report from the Public Accounts Committee, around £170 million to £410 million of taxpayer money was lost due to the failure of Metronet. The companies involved in the consortium collectively lost around £350 million in the collapse.
>
> **Metronet, Wikipedia**

Chapter 11

Review

It has been said that collaboration is a state of mind, a type of behavior, something to do with culture, or all about leadership style. We hope that we have convinced you that this is not the case. We have demonstrated that if you design a way of working (Target Operating Model) based on an organizational structure using common-sense management know-how and you include your closest partners in the loop, then you will create a business framework in which collaboration will thrive. It is crucial to treat this as a management discipline that can be repeated, audited, and improved just as you do with sales, commercial, finance, etc. It is not 'rocket science', and if you follow the step-by-step approach that we described, you can't go wrong.

The best thing about being a good collaborator is that it brings benefits that are beyond the reach of others. Before long you will have created a capability that unlocks superior competitiveness, efficiency, and profitability. Moreover, your organization will be a happier place to work so personnel at all levels will 'catch the bug' and new ideas will flow. Good staff will clamor to join you. Finally, you will inevitably gain a reputation as a 'successful collaborator' and attract top-quality partners from the market.

Go out and smell the roses (Figure 11.1)!

> We started with this partner and built our collaborative capability in-house over the years. We have since rolled out the model with others and currently it is worth around $35 million per year.
>
> **SCM Director, United Biscuits**

DOI: 10.4324/9781003221340-11

Figure 11.1 Flying in tight formation.

Index